Sediment Dredging at SUPERFUND MEGASITES

Assessing the Effectiveness

Committee on Sediment Dredging at Superfund Megasites

Board on Environmental Studies and Toxicology

Division on Earth and Life Studies

NATIONAL RESEARCH COUNCIL
OF THE NATIONAL ACADEMIES

THE NATIONAL ACADEMIES PRESS
Washington, D.C.
www.nap.edu

THE NATIONAL ACADEMIES PRESS   500 Fifth Street, NW   Washington, DC 20001

NOTICE: The project that is the subject of this report was approved by the Governing Board of the National Research Council, whose members are drawn from the councils of the National Academy of Sciences, the National Academy of Engineering, and the Institute of Medicine. The members of the committee responsible for the report were chosen for their special competences and with regard for appropriate balance.

This project was supported by Contract 68-C-03-081 between the National Academy of Sciences and the U.S. Environmental Protection Agency. Any opinions, findings, conclusions, or recommendations expressed in this publication are those of the authors and do not necessarily reflect the views of the organizations or agencies that provided support for this project.

International Standard Book Number-13: 978-0-309-10977-2 (Book)
International Standard Book Number-10: 0-309-10977-9 (Book)
International Standard Book Number-13: 978-0-309-10978-9 (PDF)
International Standard Book Number-10: 0-309-10978-7 (PDF)
Library of Congress Control Number 2007931836

Photograph of environmental dredging to remove contaminated sediments in the Fox River, Wisconsin. Courtesy of Gregory A. Hill, Wisconsin Department of Natural Resources.

Additional copies of this report are available from

The National Academies Press
500 Fifth Street, NW
Box 285
Washington, DC 20055

800-624-6242
202-334-3313 (in the Washington metropolitan area)
http://www.nap.edu

THE NATIONAL ACADEMIES
Advisers to the Nation on Science, Engineering, and Medicine

The **National Academy of Sciences** is a private, nonprofit, self-perpetuating society of distinguished scholars engaged in scientific and engineering research, dedicated to the furtherance of science and technology and to their use for the general welfare. Upon the authority of the charter granted to it by the Congress in 1863, the Academy has a mandate that requires it to advise the federal government on scientific and technical matters. Dr. Ralph J. Cicerone is president of the National Academy of Sciences.

The **National Academy of Engineering** was established in 1964, under the charter of the National Academy of Sciences, as a parallel organization of outstanding engineers. It is autonomous in its administration and in the selection of its members, sharing with the National Academy of Sciences the responsibility for advising the federal government. The National Academy of Engineering also sponsors engineering programs aimed at meeting national needs, encourages education and research, and recognizes the superior achievements of engineers. Dr. Charles M. Vest is president of the National Academy of Engineering.

The **Institute of Medicine** was established in 1970 by the National Academy of Sciences to secure the services of eminent members of appropriate professions in the examination of policy matters pertaining to the health of the public. The Institute acts under the responsibility given to the National Academy of Sciences by its congressional charter to be an adviser to the federal government and, upon its own initiative, to identify issues of medical care, research, and education. Dr. Harvey V. Fineberg is president of the Institute of Medicine.

The **National Research Council** was organized by the National Academy of Sciences in 1916 to associate the broad community of science and technology with the Academy's purposes of furthering knowledge and advising the federal government. Functioning in accordance with general policies determined by the Academy, the Council has become the principal operating agency of both the National Academy of Sciences and the National Academy of Engineering in providing services to the government, the public, and the scientific and engineering communities. The Council is administered jointly by both Academies and the Institute of Medicine. Dr. Ralph J. Cicerone and Dr. Charles M. Vest are chair and vice chair, respectively, of the National Research Council.

www.national-academies.org

COMMITTEE ON SEDIMENT DREDGING AT SUPERFUND MEGASITES

Members

CHARLES R. O'MELIA (*Chair*), Johns Hopkins University, Baltimore, MD
G. ALLEN BURTON, Wright State University, Dayton, OH
WILLIAM H. CLEMENTS, Colorado State University, Fort Collins
FRANK C. CURRIERO, Johns Hopkins Bloomberg School of Public Health, Baltimore, MD
DOMINIC DI TORO, University of Delaware, Newark
NORMAN R. FRANCINGUES, OA Systems Corporation, Vicksburg, MS
RICHARD G. LUTHY, Stanford University, Stanford, CA
PERRY L. MCCARTY, Stanford University, Stanford, CA
NANCY MUSGROVE, Management of Environmental Resources, Inc., Seattle, WA
KATHERINE N. PROBST, Resources for the Future, Washington, DC
DANNY D. REIBLE, University of Texas, Austin
LOUIS J. THIBODEAUX, Louisiana State University, Baton Rouge
DONNA J. VORHEES, The Science Collaborative, Ipswich, MA
JOHN R. WOLFE, Limno-Tech, Inc., Ann Arbor, MI

Project Staff

KARL E. GUSTAVSON, Project Director
RAYMOND A. WASSEL, Program Director
NORMAN GROSSBLATT, Senior Editor
MIRSADA KARALIC-LONCAREVIC, Manager, Technical Information Center
MORGAN MOTTO, Senior Project Assistant
RADIAH ROSE, Senior Editorial Assistant

Sponsor

U.S. ENVIRONMENTAL PROTECTION AGENCY

Pesticides in the Diets of Infants and Children (1993)
Dolphins and the Tuna Industry (1992)
Science and the National Parks (1992)
Human Exposure Assessment for Airborne Pollutants (1991)
Rethinking the Ozone Problem in Urban and Regional Air Pollution (1991)
Decline of the Sea Turtles (1990)

Copies of these reports may be ordered from the National Academies Press
(800) 624-6242 or (202) 334-3313
www.nap.edu

Preface

Contaminated sediments in aquatic environments pose health risks to humans and other organisms. Nationwide, the full extent of the problem is poorly documented, but it is well known that rivers, harbors, lakes, and estuaries fed by current or former industrial, agricultural, or mining areas frequently contain contaminated sediments. It is also well known that contaminants in the sediments can directly harm aquatic organisms or accumulate in their tissues, which can be consumed by humans. The potential adverse effects on human health and the environment are compelling reasons to reduce exposures.

From a regulatory standpoint, contaminated sediments are challenging to manage. The Superfund program, administered by the U.S. Environmental Protection Agency (EPA), is intended to protect human health and the environment from sites contaminated with hazardous substances. An array of techniques are available for remediating contaminated sediments, each with advantages and disadvantages. Decisions about which remedial measures to implement and, in particular, whether to dredge at contaminated sediment sites have proved to be among the most controversial at Superfund megasites. The scientific and technical difficulties of deciding on a remedial option are augmented by the challenges of implementing a regulatory authority that holds responsible parties liable for paying for the cleanup.

Regardless of cost or controversy, achieving the expected effect of remedial actions—improvements in the environment—is of primary importance. That is true for regulators who may require cleanup of a site, parties responsible for funding the cleanup, and communities and user groups affected by the contamination. This report, one piece of a continuing dialogue, seeks to assess the effectiveness of environmental dredging in reducing risks associated with contaminated sediments, particularly at large, complex Superfund sites (these sites are termed "megasites" when the cost of remedial activities is anticipated to exceed $50 million).

Over the course of its study, the Committee on Sediment Dredging at Superfund Megasites held three public sessions at which it heard presentations on dredging projects and received input from members of the public and other interested parties; two closed, deliberative sessions were held over the course of the year-long study. The report consists of six chapters. Chapters 1 through 3 introduce the problem, provide background on the issues, and describe the committee's approach to addressing the statement of task. Chapter 4 considers the data on various dredging sites to develop recommendations for implementing sediment-management techniques. Chapter 5 evaluates current monitoring approaches and suggests future approaches. Finally, Chapter 6 takes a broader look and considers sediment management at the national level, and it provides conclusions and recommendations to improve decision-making in the future.

The committee gratefully acknowledges the following for making presentations and for providing information during the committee's meetings: Loretta Beaumont, U.S. House of Representatives; Stephen Ells, Leah Evison, Elizabeth Southerland, Dave Dickerson, James Brown, Young Chang, William "Skip" Nelson, and Marc Greenberg, U.S. Environmental Protection Agency; Michael Palermo, U.S. Army Corps of Engineers (retired); Clay Patmont, Anchor Environmental; Steven Nadeau, Sediment Management Work Group; John Connolly, QEA; Larry McShea, Alcoa; Rick Fox, Natural Resource Technology; Mike Jury, CH2M Hill; John Kern, KERN Statistical Services; and Todd Bridges, U.S. Army Corps of Engineers.

The committee is also grateful for the assistance of National Research Council staff in preparing this report: Karl Gustavson, study director; James Reisa, director of the Board on Environmental Studies and Toxicology; Ray Wassel, program director; Norman Grossblatt and Ruth Crossgrove, senior editors; Mirsada Karalic-Loncarevic, manager of the Technical Information Center; Morgan Motto, senior project assistant; and Radiah Rose, senior editorial assistant. Finally, I thank the members of the committee for their dedicated efforts throughout the development of this report.

Charles R. O'Melia, *Chair*
Committee on Sediment Dredging at
Superfund Megasites

Acknowledgment of Review Participants

This report has been reviewed in draft form by persons chosen for their diverse perspectives and technical expertise in accordance with procedures approved by the National Research Council (NRC) Report Review Committee. The purpose of this independent review is to provide candid and critical comments that will assist the institution in making its published report as sound as possible and to ensure that the report meets institutional standards of objectivity, evidence, and responsiveness to the study charge. The review comments and draft manuscript remain confidential to protect the integrity of the deliberative process. We wish to thank the following for their review of this report:

Todd S. Bridges, U.S. Army Engineer Research and Development Center
Edwin H. Clark II, Earth Policy Institute
J. Paul Doody, Blasland, Bouck & Lee, Inc., an Arcadis Company
Rick Fox, Natural Resource Technology, Inc.
Paul Fuglevand, Dalton, Olmsted & Fuglevand, Inc.
John C. Henningson, Henningson Environmental Services
Chris Ingersoll, U.S. Geological Survey
Stephen U. Lester, Center for Health & Environmental Justice
Jeffrey S. Levinton, Stony Brook University
Victor S. Magar, ENVIRON
Larry McShea, Alcoa
Steven C. Nadeau, Honigman Miller Schwartz and Cohn, LLP
Clay Patmont, Anchor Environmental, LLC.
Harlee S. Strauss, H. Strauss Associates, Inc.

Although the reviewers listed above have provided many constructive comments and suggestions, they were not asked to endorse the conclusions or recommendations, nor did they see the final draft of the report before its release. The review of this report was overseen by George M. Hornberger, University of Virginia, and C. Herb Ward, Rice University. Appointed by the NRC, they were responsible for making certain that an independent examination of this report was carried out in accordance with institutional procedures and that all review comments were carefully considered. Responsibility for the final content of this report rests entirely with the author committee and the institution.

Abbreviations

AOCs	areas of concern
ARARs	applicable or relevant and appropriate requirements
ARCS	Assessment and Remediation of Contaminated Sediments
AVS	acid volatile sulfide
BMPs	best management practices
CFR	Code of Federal Regulations
COCs	contaminants of concern
COPCs	contaminants of potential concern
CSO	combined sewer outfalls
DGPs	differential global positioning systems
ELISA	enzyme-linked immunosorbent assays
EPA	U.S. Environmental Protection Agency
EQC	environmental quality criteria
KDGPS	kinematic differential global positioning systems
MC	Main Channel
MCSS	Major Contaminated Sediment Sites Database
MNR	monitored natural recovery
MTCA	Model Toxics Control Act
NHANES	National Health and Nutrition Examination Survey
NNS	Northern Near Shore
NOAA	National Oceanographic and Atmospheric Administration
NPL	National Priorities List
NSI	National Sediment Inventory
OMC	Outboard Marine Corporation
PAHs	polycyclic aromatic hydrocarbons
PCBs	polychlorinated biphenyls
PED	polyethylene device
PNEC	probable no effect concentration
PRA	probabilistic risk assessment
RAL	remedial action level
RI/FS	remedial investigation and feasibility study
ROD	record of decision

ROPS	remedial options pilot study
SMA	sediment management area
SMS	Sediment Management Standards
SMU	Sediment Management Unit
SPI	sediment profile imagery
SPMD	semipermeable membrane device
SPME	solid-phase microextraction
SQG	sediment quality guidelines
SQO	sediment quality objectives
TSCA	Toxic Substances Control Act
UCL	upper confidence limit

Contents

APPENDIXES

BOXES, TABLES, AND FIGURES

BOXES

FIGURES

TABLES

Sediment Dredging at SUPERFUND MEGASITES

Assessing the Effectiveness

Summary

BACKGROUND

Contaminated sediments in aquatic environments can pose risks to human health and other organisms. Nationwide, the full extent of the problem is poorly documented, but it is well known that many rivers, harbors, lakes, and estuaries fed by existing or former industrial, agricultural, or mining areas contain contaminated sediments. It is also well known that contaminants in the sediments can directly harm aquatic organisms or accumulate in their tissue, which can be consumed by humans. The potential adverse effects on human health and the environment are compelling reasons to reduce such exposures.

From a regulatory standpoint, contaminated sediments are challenging to manage. The Superfund program,[1] administered by the U.S. Environmental Protection Agency (EPA), is intended to protect human health and the environment from hazardous substances at contaminated sites. At most contaminated sediment Superfund sites, the remedial

---

[1]The Superfund requirements are set forth in the Comprehensive Environmental Response, Compensation, and Liability Act (CERCLA) of 1980 as amended (42 USC §§ 9601-9675 [2001]), and its implementing regulations are set forth in the National Contingency Plan (NCP) (40 CFR 300).

process includes a site investigation, comparison of remedial alternatives, and selection and implementation of a remedy. The process is affected by numerous scientific and technical issues. Resource-intensive surveys and analyses are required to document the distribution, depth, and concentration of contaminants in sediments and contaminant concentrations in the aquatic biota. Even if substantial resources are focused on a relatively small site, understanding the current and potential risks of contaminated sediments can be difficult and uncertain, relying heavily on surrogate measures and modeling of actual environmental effects. The process of estimating and comparing the modeled results of potential remedial actions (including no action) has substantial uncertainties that depend on a host of variables, including whether environmental conditions have been adequately characterized and the accuracy of near-term and long-term predictions of post-remediation contaminant behavior. The uncertainties are magnified by increasing duration of a remedial action and increased extent and complexity of a contaminated site.

Contaminated sediments exist in a variety of environments and can differ greatly in type and degree of contamination. Site conditions are important in determining which remediation techniques (and combinations thereof) are appropriate. The techniques include removing the sediments from the aquatic environment (for example, by dredging), capping or covering contaminated sediments with clean material, and relying on natural processes while monitoring the sediments to ensure that contaminant exposures are decreasing, or at least not increasing. Those approaches differ in complexity and cost; dredging is the most complex and expensive, and monitoring without active remediation is the least difficult and least expensive. Remedial approaches have trade-offs with respect to the risks that are created during implementation and that remain after remediation. Dredging may create exposures (for example, through the resuspension of buried contaminants) during implementation, but it has the potential to remove persistent contaminants permanently from the aquatic environment. Monitoring without removal does not itself create risks, but it leaves contaminants in the aquatic environment. Remedial operations also vary in efficacy within and among the different approaches. The variability is driven by several factors, including site conditions and implementation of the remedial approach.

Decisions about whether to dredge at contaminated sediment sites have proved to be among the most controversial at Superfund megasites.[2] The scientific and technical difficulties described above are augmented by the challenges of implementing a program that holds responsible parties liable for paying for the cleanup, with parties in some cases unwilling to accept the liability. Cleanup planning must also be responsive to the public, which may have little tolerance for remedial actions that leave contaminants in the local environment. Those controversies often expand with the magnitude of the sites and the scope of remedial activity, as has been seen at some of the nation's largest sediment remediation sites (for example, the Fox River, WI, and the Hudson River, NY).

Regardless of cost or controversy, achieving the intended effect of remedial actions in terms of anticipated improvements in the environment is of primary importance. That is true for regulators who require cleanup of a site, parties responsible for funding the cleanup, and communities and user groups, such as anglers and boaters, who are directly affected by the contamination.

THE COMMITTEE'S CHARGE

This study, which was conducted to evaluate the effectiveness of dredging as a remedial option at contaminated sediment sites, originated in the fiscal year 2006 appropriations bill for the Department of Interior, environment, and related agencies. The accompanying conference report (Report 109-188) states that "the managers believe that the appropriate role for the NAS [National Academy of Sciences] is to act as an independent peer review body that will conduct an objective evaluation of some of the ongoing dredging projects underway at Superfund megasites. By undertaking such an evaluation, the Academy can serve as an objective voice on this issue."

The Committee on Sediment Dredging at Superfund Megasites was convened by the National Research Council of the National Academies. In brief, the committee's charge requests an evaluation of the expected effectiveness of dredging of contaminated sediments at Superfund megasites. The committee was asked to consider such aspects of dredging as short-term and long-term changes in contaminant transport and ecologic effects. Overall, the committee was charged to strive to develop

---

[2]Megasites are those Superfund sites where remedial expenditures are expected to exceed $50 million.

recommendations that would facilitate scientifically based and timely decision-making for megasites in the future but not to recommend particular remedial strategies for specific sites.

The committee, chosen by the National Research Council, consists of experts in a variety of fields relevant to the statement of task. Over the course of the study, the committee held three public sessions in which it heard presentations on an array of dredging projects and received input from members of the public and other interested parties; two closed, deliberative sessions were also held over the course of the year-long study.

EVALUATION OF EXPERIENCE AT DREDGING SITES

The committee examined experience at 26 dredging projects and evaluated whether, after dredging, the cleanup goals had been met. That involved evaluating whether a site achieved its quantitative cleanup levels (typically a specified concentration in sediment expected to be achieved in the short term, that is, immediately after remedy implementation) and remedial-action objectives (often a narrative statement of what the cleanup is expected to accomplish in the long term).[3] Various sites were examined, including full-scale dredging projects, pilot studies at sites, and dredging projects within a large-scale remediation effort.

Conclusions

The committee concluded that dredging is one of the few options available for the remediation of contaminated sediment and that it should be considered, with other options, to manage the risks that the contaminated sediments pose. However, the committee could not generally establish whether dredging *alone* is capable of long-term risk reduction. That is because monitoring at most sites does not include all the

---

[3]In this context, *short term* refers to anything caused as an immediate consequence of the action being focused on; *long-term* extends beyond this time period to the time required to achieve remedy success. For dredging projects, this would include any period of natural recovery that is part of the remedy and necessary to achieve the goals of the remedy.

BOX S-1 Overview of Conclusions and Recommendations

Sediment dredging can be effectively implemented to remove contaminants from aquatic systems, but technical limitations often constrain its ability to achieve expected outcomes. The range of experiences and outcomes at dredging sites, coupled with shortcomings in monitoring data, the lack of sufficient time to observe long-term changes, and difficulties in separating the effects of dredging from the effects of other processes limited the committee's ability to establish whether dredging *alone* has been effective in risk reduction. However, assessment of data from dredging projects does indicate that dredging has encountered systematic difficulties in achieving specified cleanup levels (expected sediment-contaminant concentrations after dredging) and that monitoring to evaluate long-term success is generally lacking. The inability to meet desired cleanup levels is associated primarily with "residual" contamination that typically results from dredging operations or from leaving contaminated sediment exposed after dredging. Site assessments also indicate that contaminants can be released into the water during dredging and can have short-term adverse effects on the aquatic biota. Residual contamination and contaminant release are inevitable during dredging and should be explicitly considered in estimating risk reduction. Some site conditions and dredging practices can limit the amount of residual contamination remaining after dredging and can limit contaminants released into the water column. Those site conditions should be given major consideration when evaluating the potential effectiveness of dredging.

Environmental monitoring is the only way to evaluate remedial success, but monitoring at most Superfund sites has been inadequate to determine whether dredging has been effective in achieving remedial objectives (that has not been the case in several highly monitored pilot studies). Basic information was not collected at some sites, and others had only recently completed dredging, so long-term trends could not be assessed. EPA should ensure that adequate monitoring is conducted at all contaminated sediment megasites to evaluate remedial effectiveness. Some current monitoring techniques have proved useful in determining short-term and long-term effects of remediation, but further development of monitoring strategies is needed. Pre-remediation monitoring is necessary to adequately characterize site conditions and to assemble a consistent long-term dataset that allows statistically valid comparisons with future post-remediation monitoring data. Monitoring data should also be made available to the public in an accessible electronic form so that evaluations of remedial effectiveness can be independently verified.

(Continued on next page)

BOX S-1 Continued

Regarding the future practices and management of contaminated sediments at megasites, the committee recommends that adaptive-management approaches should be implemented in the selection and implementation of remedies where there is a high degree of uncertainty about the effectiveness of the remedial action. In selecting site remedies, dredging remains one of the few options available for the remediation of contaminated sediment, and it should be considered along with other options, but EPA should compare the expected net risk reduction associated with each remedial alternative, taking into account the range of risks and uncertainty associated with each alternative. EPA should centralize sediment-related resources, responsibility, and authority at the national level to ensure that necessary improvements are made in site tracking, in the implementation of monitoring and adaptive management, and in research to examine the relationship between the remedial actions, site conditions, and risk reduction.

---

measures necessary to evaluate risk over time,[4] dredging may have occurred in concert with other remedies or natural processes that affect risk, insufficient time has passed to evaluate long-term risk reduction, and a systematic compilation of site data necessary to track remedial effectiveness nationally is lacking.

However, the committee was able to draw several conclusions and derive recommendations on the basis of monitoring data from a range of dredging projects and by evaluating factors that affected their success. The analysis indicates that dredging can be effective for removal of mass, but that mass removal alone does not necessarily achieve risk-based goals. Monitoring data demonstrate that dredging can have short-term adverse effects, including increased contaminant concentrations in the water, increased contaminant concentrations in the tissues of caged fish adjacent to the dredging activity, and short-term increases in tissue contaminant concentrations in other resident biota. However, monitoring for those effects was not conducted at many sites.

---

[4]Monitoring for risk reduction is not straightforward; there is no analytic determination of "risk," and estimating risk reduction requires multiple metrics (for example, concentration, toxicity and bioavailability data) to be measured consistently through time.

The most frequent post-dredging measurement used to assess effectiveness at the sites was contaminant concentrations in surface sediment. Surface concentrations (as opposed to concentrations in deeply buried sediments) are the most relevant to risk. At some sites (for example, Bayou Bonfouca, LA; Waukegan Harbor, IL; and the Dupont Newport Site, Christina River, DE), sediments were not sampled for contaminants immediately following dredging. The committee's analysis of pre-dredging and post-dredging surface sediment concentrations indicates a wide range of outcomes: some sites showed increases, some no change, and some decreases in contaminant concentrations. Residual contamination after dredging can result from the incomplete removal of targeted sediments or the deposition of sediment resuspended during dredging. Residual contaminated sediments hamper the ability to achieve desired cleanup levels and are exacerbated by site conditions like obstructions in the dredging area and impenetrable or uneven formations underlying the contaminated sediments. Overall, the committee found that dredging alone achieved the desired contaminant-specific cleanup levels at only a few of the 26 dredging projects, and that capping[5] after dredging was often necessary to achieve cleanup levels.

The committee was able to identify factors that led to the success or failure of projects to meet desired short-term cleanup levels and, where long-term data were available, remedial-action objectives. Some sites exhibited conditions that are more conducive to dredging and less prone to releasing contaminants and less likely to result in residual contaminated sediment after dredging. Favorable site conditions include little or no debris (for example, rocks, boulders, cables, automobiles, and I-beams), sediment characteristics that permit rapid (even visual) determination of clean vs contaminated sediment, conditions that allow over-dredging into clean material beneath contaminated sediment (sites underlain by bedrock or hardpan are highly problematic), low-gradient bottom and side slopes, lack of piers and other structures, rapid natural attenuation processes after dredging, and absence of contaminants that distribute to the water column rapidly after sediment disturbance.

---

[5]*Capping* refers broadly to the placement of a layer of uncontaminated material over material with elevated concentrations to contain contaminated sediment.

The design and implementation of remediation can also influence the extent of chemical release and residual contamination (as well as counteract the influence of poor site conditions). Adequate site characterization is needed to identify adverse conditions and potential sources of recontamination in the site or watershed. Pilot studies are particularly useful for identifying adverse site conditions and logistical problems. As discussed in the report, best-management practices during dredging can help control residuals and resuspension. Backfilling and capping can be used following dredging to manage residual contaminated sediments. Contracting mechanisms can be used to encourage a focus on specified cleanup levels and remedial-action objectives instead of on attaining mass removal targets.

The combined experience indicates that dredging *alone* is unlikely to be effective in meeting short-term and long-term goals if a site has one or more unfavorable conditions. If conditions are unfavorable, contaminant resuspension and release and residual contamination will tend to limit the ability to meet desired cleanup levels and will delay the achievement of remedial-action objectives unless managed with a combination of remedies.

Recommendations

• Remedies should be designed to meet long-term risk-reduction goals (as opposed to metrics not strictly related to risk, such as mass-removal targets). The design should be tested by modeling and monitoring the achievement of long-term remedial action objectives.

• Environmental conditions that limit or favor the effectiveness of dredging should be given major consideration in deciding whether to dredge at a site.

• Resuspension, release, and residuals will occur if dredging is performed. Decision-making should include forecasts to estimate the effects of those processes, and the predictions should be explicitly considered in expectations of risk reduction. To reduce adverse effects, best-management practices that limit resuspension and residual contamination should be used during dredging. The ability of combination remedies to lessen the adverse effect of residuals should be considered.

- Further research should be conducted to define mechanisms, rates, and effects of residuals and contaminant resuspension associated with dredging. It is known that contaminated sediment resuspension and residuals create exposures, but the relationship of the magnitude of those processes to environmental conditions, operational controls, and management practices is not well quantified.

MONITORING FOR EFFECTIVENESS

Conclusions

Environmental monitoring is the only way to evaluate a remedy's success in reducing risk and ensure that the objectives of remediation have been met. It is therefore an essential part of the remedy. It is impossible to evaluate effectiveness in the absence of sufficient baseline data and appropriate reference sites. Monitoring needs to be conducted to confirm not only that desired cleanup levels have been met, but that they result in risk reduction.

In most cases reviewed by the committee, monitoring was designed or implemented inadequately to determine whether dredging was effective in achieving the objectives of the remediation or long-term risk reduction. For example, at some sites, sparse or incomplete monitoring data were collected. Pre-remediation monitoring approaches were not always consistent with those used for long-term post-remediation monitoring. Pre-remediation trends in sediment or fish concentrations were not of sufficient duration to enable judging the effect of the remedial action. The models and forecasts used to select a remedy were not updated with post-remediation data to determine whether remediation had the expected effects (or to examine why or why not). Monitoring was of insufficient quality or quantity to support rigorous statistical analyses.

However, some monitoring techniques have proved useful in determining short-term and long-term effects of remediation. Monitoring during dredging, including measurements of mass flux (contaminant transport over time) attained through upstream and downstream chemical monitoring, and biologic monitoring, including caged-fish studies and passive sampling devices,[6] are useful in indicating chemical releases to the water from dredging. Chemical concentrations do not always correspond directly with potential uptake or toxicity because contaminant bioavailability can differ among sampling locations. However, chemical analyses are among the most highly standardized and easily attainable indicators of risk and are useful in evaluating trends before and after dredging. Laboratory toxicity testing has proven valuable in long-term monitoring of risk to benthic organisms in sediment following dredging.

Fish-tissue monitoring for contaminants is useful for indicating risks to the health of people and other piscivorous species, particularly if long-term monitoring trend data exist. However, linking changes in fish-tissue concentrations to remedial actions can be problematic, because fish are mobile and can be exposed to offsite conditions. For describing possible ecologic effects, benthic organisms (or organisms with home ranges limited to the site) or passive sampling devices are probably better indicators, although not necessarily sufficient indicators of exposure to higher trophic level species.

Recommendations

- EPA should ensure that monitoring is conducted at all contaminated sediment megasites to evaluate remedy effectiveness. Monitoring data should be made available to the public in a form that makes it possible to verify evaluations of remedial efficacy independently.
- Monitoring plans should focus on elements required to judge remedial effectiveness and to inform management decisions about a site. Planning, evaluation, and adaptive management[7] based on monitoring findings should be closely linked to the conceptual site model so that the hypotheses and assumptions that led to the selected remedy can be tested and refined.
- Pre-remediation baseline monitoring methods and strategies should be developed to allow statistically valid comparisons with post-

---

[6]Passive sampling devices accumulate chemicals of interest during an extended deployment period (days to weeks) in the environment to provide an integrated estimate of chemical exposure over that period.

[7]Adaptive management is generally used to learn from, test, assess, and modify or improve remedies with the goal of meeting long-term objectives.

remediation monitoring datasets. The ultimate goal is to assemble a consistent long-term dataset that can be used in evaluations. Monitoring should be initiated during the design of the remedy to help to establish a pre-remedial time trend, integrating earlier characterization data as technically appropriate.

- Research in and development of rapid field monitoring techniques to inform dredging operations in nearly real time are needed to indicate the effects of resuspension of contaminants and their release to the water column. Biota monitoring approaches that use benthic invertebrates (or other organisms with appropriate home ranges) as indicators of food-web transfer of contaminants should also be developed.

IMPROVING FUTURE DECISION-MAKING AT SUPERFUND MEGASITES

Conclusions

The historical perspective and hindsight gained from the committee's retrospective analysis of sediment sites provide an opportunity to derive common lessons and to improve on the manner in which environmental dredging is planned and implemented. It is important that this type of review be on-going and be part of a shared experience among regulators, practitioners, and the public. The large spatial scale and long remedial timeframes of contaminated sediment megasites make it difficult to predict and quantify the human health and ecosystem risk-reduction benefits achieved by isolated remediations in a large-scale watershed. In addition, the complexity and heterogeneity of large-scale megasites suggest that a variety and combination of remedial approaches will often be appropriate. The committee concludes that three critical kinds of changes are needed to improve decision-making and the efficacy of dredging remedies at contaminated sediment megasites.

First, owing to the complexity, large spatial scale, and long time frame involved, the management of contaminated megasites needs to embrace a more flexible and adaptive approach to accommodate unanticipated factors, new knowledge, technology changes, and results of field pilot tests. Comprehensively characterizing sediment deposits and contaminant sources on the scale of megasites presents tremendous

technical challenges and uncertainties. Many large and complex contaminated sediment sites will take years or decades to remediate and will
encounter unforeseen conditions. A priori predictions of the outcome of
that remediation, made on the basis of incomplete information, will also
have high uncertainty. The typical Superfund approach, wherein EPA
conducts a remedial investigation and a feasibility study that establishes
a single path to remediation in the record of decision is not the best approach to remedy selection and implementation at these sites. At the
largest contaminated sediment sites, the remediation timeframes and
spatial scales are in many ways unprecedented. Remedial strategies will
often require unexpected adjustments, whether in response to new
knowledge about site conditions or advances in technology (such as improved dredge or cap design or in situ treatments). Because such uncertainty exists, regulators and others will need to adapt continuously to
evolving conditions and environmental responses that cannot be foreseen. Thus, the process for remedy selection at large, complex sediment
megasites needs to allow more adaptive site investigation, remedy selection, and remedy implementation.

Second, improved risk assessment that specifically considers the
full range and real-world limitations of remedial alternatives is needed
to allow valid comparisons of technologies and uncertainties. Each remedial alternative offers a unique set of risk-reduction benefits, possibly
with the creation of new contaminant exposure pathways and associated
risks. The effects of adverse environmental conditions, such as those
leading to chemical release and production of contaminated residuals,
need to be accounted for in a quantitative comparison of net risk reduction associated with different alternatives, and uncertainty should be
quantified to the extent warranted to optimize decision-making. Some
potential effects not related to chemical exposures (for example, quality
of life impacts and some risks to workers or the community from implementing a remedy) will remain difficult to quantify and to compare.
However, ignoring these risks in comparisons of remedial alternatives is
not the solution and may lead to undesirable consequences. Quantitative
assessment and comparison of some of these types of risks will probably
never be fully achievable, but they should be identified as risks associated with particular alternatives and considered at least qualitatively
when remedial alternatives are being compared and the risks and benefits associated with various options are presented to the public.

Third, EPA needs to centralize and coordinate assessment and management of contaminated sediment megasites to ensure greater consistency in evaluations, greater technical competence, more active leadership at the sites, and an emphasis on what works and why. Several specific responsibilities are discussed in the corresponding recommendation below.

Recommendations

•   An adaptive-management approach is essential to the selection and implementation of remedies at contaminated sediment megasites where there is a high degree of uncertainty about the effectiveness of dredging.

•   Adaptive approaches based on the use of monitoring data from pilot studies and remedial operations should be used to learn from, test, assess, and modify or improve remedies with the goal of meeting long-term objectives.

•   EPA should compare the estimated net risk reductions associated with different remedial alternatives, taking into account the real-world limitations of each approach (such as residuals and resuspension) in selecting site remedies.

•   EPA should centralize resources, responsibility, and authority at the national level to ensure that necessary improvements are made so that contaminated sediment megasites are remediated as effectively as possible. Responsibilities would include

— *Gathering data to define the scope of the contaminated sediment problem nationally and track likely future contaminated sediment megasites.* Defining the scope of the contaminated sediment problem is important for two reasons: this will help place the magnitude of the problem in proper perspective by establishing how much of the problem has been addressed and how much remains, and documenting remaining work and associated costs should help EPA and Congress identify the most pressing program and research needs.

— *Reviewing site studies, remedies, and monitoring approaches at contaminated sediment megasites to assess whether best practices are being implemented, including whether regions are complying with national sedi-*

ment and other program guidance. Because every EPA region has on-the-ground management and remediation experience with dredging at some megasites, regular review and shared experience can inform decision-making and raise the overall level of technical expertise. The goal is to generate a greater understanding of sound remediation principles and best practices and their consistent application among sites.

— *Ensuring that adaptive-management principles and approaches are applied at contaminated sediment megasites.* As described above, a phased, adaptive approach will be required in remedy selection and implementation at large, complex megasites. EPA should ensure that monitoring data are used to support and update forecasts of the effects of remedial measures and to adapt a remedy if remedial goals are not achieved in the expected time frame.

— *Ensuring adequate pre-remediation and post-remediation monitoring and evaluating sediment cleanups in nearly real time to determine whether remedies are having the intended effects.* Without adequate pre-dredging and post-dredging monitoring, it is impossible to evaluate the degree to which cleanup has achieved remedial objectives. EPA should invest in better and more consistent measurement tools to monitor conditions in the field reliably and efficiently. EPA should ensure that these techniques are used before and after remediation so that the effectiveness of the projects can be assessed. To facilitate information transfer, a centralized, easily accessible, and up-to-date repository of relevant data and lessons learned regarding sediment remedies should be created.

— *Developing and implementing a research strategy, including new technologies, for contaminated sediment sites.* EPA and its federal partners should develop a research and evaluation strategy to understand the risk reduction attained by various technologies under various site conditions and the associated uncertainty. EPA's efforts should focus on moving forward with remedies at sites while testing and learning with each new pilot test or remedy to determine what works, what does not work, and why. Research to improve and develop new remediation technologies, site-characterization techniques, and monitoring tools is essential to advance sediment remediation and should be supported.

Many of the sites are vast and expensive, and it is worthwhile to invest time and resources now to ensure more cost-effective remedies in the future. That focus is warranted if the country is to make the best use of the billions of dollars yet to be spent on remediation.

1

Introduction

THE CHALLENGE OF CONTAMINATED SEDIMENTS

Contaminated sediments in aquatic environments can pose health risks to many types of organisms, including humans. Exposure to the contaminants occurs by several routes, including direct contact and consumption of organisms that have accumulated contaminants from the sediments. The potential adverse effects on human health and the environment are compelling reasons to seek to reduce exposure.

Contaminated sediments can occur in small, localized areas or in vast areas, covering miles of river or harbor bottoms and associated floodplains. They occur in wetlands, coastal tidal flats and embayments, ocean basins, lakes, rivers, and streams. In some cases, contamination is relatively contained; in other cases, contaminated sediment exists throughout a watershed and may have multiple sources of contamination, including stormwater and sewer outfalls, industrial discharges, agricultural runoff, and atmospheric deposition.

The chemicals of concern in contaminated sediment sites vary; polychlorinated biphenyls (PCBs) are the most common, followed by metals, and polycyclic aromatic hydrocarbons (EPA 2005). The widely varied physical and chemical properties of contaminants markedly affect their distribution in the environment and their behavior (including transport, bioavailability, and toxicity) during and after remediation. The degree of contamination can be severe in some areas with nearly unadulterated original products, such as PCB-containing oils, pesticides, or coal-tar residues. In other areas, contaminants occur at low concentrations in sediments among functioning ecosystems of fish, plants, and benthic invertebrates. The thickness of the contaminated sediment is highly variable and often poorly characterized but can range from a few inches to many feet thick with marked differences over small spatial scales. In addition, the nature of the sediments and particularly of the underlying substrate can vary widely on the basis of local geology, hydrology, and human activities that have altered the watershed characteristics.

Because of the highly variable nature of sediments, the environments in which they occur, and the type and degree of contamination, there are many approaches to their remediation. The techniques, which can be employed in combination, include removing the sediments from the aquatic environment (for example, by dredging), capping or covering contaminated sediments with clean material, and relying on natural processes while monitoring the sediments to ensure that contaminant exposures are decreasing, or at least not increasing (known as monitored natural recovery [MNR]). In-situ treatments that, for example, reduce the bioavailability of contaminants can also be used. The techniques, which are examined in greater detail in Chapter 2, differ in complexity, cost, efficacy, and time frame. That variability is driven by several factors, including site conditions (for example, variations in water flow and depth), underlying substrate characteristics, and implementation of the remedial approach. Regardless, achieving expected reduction in risk is of primary importance to regulators who require cleanup of a contaminated sediment site, parties responsible for funding the cleanup, and communities and user groups that are directly affected by the contamination and the remediation process.

Managing the risks associated with contaminated sediments has been an issue at the federal level since at least the middle 1970s (Johanson and Johnson 1976, as cited in EPA 1987), although it received substantially greater attention in the 1980s when the U.S. Environmental Protection Agency (EPA) sought to document the nature and extent of sediment contamination (Bolton et al. 1985; EPA 1987). The 1989 National Research Council report *Contaminated Marine Sediments: Assessment and Remediation* (NRC 1989) examined the extent of and corresponding risk posed by marine sediment contamination and examined remedial technologies. EPA's Assessment and Remediation of Contaminated Sediments (ARCS) program—an early effort to understand the extent of, associated risks of, and techniques for remediating contaminated sediments—published several useful reports and guidance documents dealing with the assessment of contaminated sediments and various treatment technologies (EPA 1994). Since then, additional National Research Council reports on managing contaminated sediments have been released (NRC 1997, 2001), and EPA has published its sediment quality surveys (EPA 1997; 2004), produced a contaminated sediment management strategy (EPA 1998), and issued comprehensive contaminated sediment guidance (EPA 2005). Yet, even after decades of analysis and review, managing and remediating contaminated sediments remains a major scientific and management challenge. Areas with contaminated sediments continue to be identified, and remediation efforts are increasingly large, expensive, and resource-intensive.

This report is one piece of the continuing dialogue and seeks to assess the effectiveness of environmental dredging for reducing risks associated with contaminated sediments, particularly at large, complex sites. Environmental dredging is of special interest because it can be expensive and technically challenging to implement. Dredging itself may create exposures (for example, through the resuspension of buried contaminants), but it removes persistent contaminants (and their associated potential for transport and risk) from the aquatic environment permanently. Whether to dredge contaminated sediments has proved to be one of the most controversial aspects of decision-making at sediment remediation sites.

THE CHARGE TO THE COMMITTEE ON SEDIMENT DREDGING AT SUPERFUND MEGASITES

This study was requested in the fiscal year 2006 appropriations bill for the Department of Interior, environment, and related agencies. The accompanying conference report (Report 109-188) states that "the managers believe that the appropriate role for the NAS [National Academy of Sciences] is to act as an independent peer review body that will conduct an objective evaluation of some of the ongoing dredging projects underway at Superfund megasites. By undertaking such an evaluation the NAS can serve as an objective voice on this issue."[1]

In response, the National Research Council of the National Academies convened the Committee on Sediment Dredging at Superfund Megasites to consider the specific tasks provided in the statement of task (see Appendix A). In brief, the committee's charge requests an evaluation of the expected effectiveness of dredging of contaminated sediments at Superfund megasites and of whether risk-reduction benefits are expected to be achieved in the expected period. The committee was asked to consider such aspects of dredging as the short- and long-term changes in contaminant transport and ecologic effects. The statement of task also directs the committee to evaluate monitoring strategies and whether those strategies are sufficient to inform assessments of effectiveness. Overall, the committee was charged to strive to develop recommendations that would facilitate scientifically based and timely decision-making for megasites in the future but not to recommend particular remedial strategies at specific sites.

One subject of great interest and concern at contaminated sediment Superfund sites is the risk-based comparison of remedial alternatives and selection of a remedy (Bridges et al. 2006; Wenning et al. 2006; Zeller and Cushing 2006). The committee briefly discusses this topic (Chapter 2) and addresses it in the context of improving future decision making at Superfund megasites (Chapter 6). However, the report does not develop specific procedures and recommendations for performing comparative risk analyses in selection of a sediment remedy. While that topic and type of analysis is quite important, it was not requested of the committee and it has not been undertaken.

NATIONAL RESEARCH COUNCIL AND THE COMMITTEE PROCESS

The National Research Council is a nonfederal, nonprofit institution that provides objective science, technology, and health-policy advice generally by producing consensus reports written by committees. It exists to provide independent advice; it has no government affiliation and

---

[1]Megasites are those Superfund sites where remedial expenditures are expected to exceed $50 million.

no regulatory role. There is no direct oversight of a committee by the study sponsor or any other outside parties. Thus, EPA and other interested parties have no more input or access to committee deliberations than does the general public. That arrangement gives the committee complete independence in conducting its study. The committee members have a wide variety of backgrounds and expertise. Members are selected by the Research Council primarily for their academic credentials and their knowledge, training, and experience relevant to the statement of task (see Appendix B for committee-member biosketches). Members conduct their work solely as a public service, volunteering to the Research Council and the nation, cognizant of the importance of providing timely and objective scientific advice.

In conducting its review and evaluation, the committee relied on the Superfund-site decision documents and supporting materials, other scientific studies, technical presentations made to the committee, other information submitted by individuals and interest groups, and the committee's observations and personal expertise. All information received by Research Council staff that was made available to committee members is available to the public through the Research Council's public-access records office.

The committee held five meetings. Three included open, information-gathering sessions in which the committee heard from invited speakers and from interested members of the public. The first meeting (in March 2006) was in Washington, DC; the second was held in Irvine, CA (June 2006); and the third was in Woods Hole and New Bedford, MA (July 2006), where the committee toured an active dredge site and sediment-handling facility. All of the public meetings included an open session where anyone was able to provide comment to the assembled committee. In addition, the committee was available to receive written materials throughout the study. The fourth and fifth meetings, held in September and October 2006 in Washington, DC, were closed, deliberative sessions attended only by committee members and staff.

REPORT ORGANIZATION

Chapter 2 provides background on sediment management at Superfund megasites; it includes discussion of the concept of reducing risk through environmental remediation and details on remedial techniques, particularly dredging. Chapter 3 describes the committee's approach to considering effectiveness at various sites and developing conclusions from the analyses. Chapter 4 evaluates remedy performance and risk reduction on the basis of sites' pre-dredging and post-dredging monitoring data and evaluates factors that affected performance. Chapter 5 looks at current practices for monitoring effectiveness at sediment remediation sites and considers the types of assessments and protocols that are needed to improve monitoring. Chapter 6 looks to the future: it considers the implications of the committee's assessment of sediment management and identifies opportunities to advance the understanding of dredging and its effectiveness in improving the environment and public health.

Overall, the committee recognizes that the state of the science of environmental dredging is continually changing. New information is being gathered, research detailing the effects and effectiveness of dredging is being conducted, and technologies and performance continue to evolve. That process will continue for the foreseeable future. The committee does not consider its review to be the last word, but it hopes that its findings and recommendations will assist government agencies and other stakeholders in improving the approaches to contaminated sediments at large, complex megasites.

REFERENCES

Bolton, H.S., R.J. Breteler, B.W. Vigon, J.A. Scanlon, and S.L. Clark. 1985. National Perspective on Sediment Quality. EPA Contract No. 68-01-6986. U.S. Environmental Protection Agency, Criteria and Standards Division, Office of Water Regulations and Standards, Washington, DC.

Bridges, T.S., S.E. Apitz, L. Evison, K. Keckler, M. Logan, S. Nadeau, and R.J. Wenning. 2006. Risk-based decision making to manage contaminated sediments. Integr. Environ. Assess. Manage. 2(1):51-58.

EPA (U.S. Environmental Protection Agency). 1987. An Overview of Sediment Quality in the United States. EPA-905/9-88-002. Office of Water Regulations and Standards, U.S. Environmental Protection Agency, Washington, DC, and Region 5, Chicago, IL. June 1987.

EPA (U.S. Environmental Protection Agency). 1994. Assessment and Remediation of Contaminated Sediments (ARCS) Program: Final Summary Report.

EPA 905-S-94-001. Great Lakes National Program Office, U.S. Environmental Protection Agency, Chicago, IL. August 1994.

EPA (U.S. Environmental Protection Agency). 1997. The Incidence and Severity
of Sediment Contamination in Surface Waters of the United States: Report
to Congress, Vols. 1-3. EPA 823-R-97-006, EPA 823-R-97-007, EPA 823-R-97-
008. Office of Science and Technology, U.S. Environmental Protection
Agency, Washington, DC. September 1997 [online]. Available: http://
www.epa.gov/waterscience/cs/congress.html [accessed Dec. 26, 2006].

EPA (U.S. Environmental Protection Agency). 1998. EPA's Contaminated Sediment Management Strategy. EPA-823-R-98-001. Office of Water, U.S. Environmental Protection Agency. April 1998 [online]. Available: http://www.
epa.gov/waterscience/cs/strategy.pdf [accessed Dec. 26, 2006].

EPA (U.S. Environmental Protection Agency). 2004. The Incidence and Severity
of Sediment Contamination in Surface Waters of the United States:
National Sediment Quality Survey, 2nd Ed. EPA-823-R-04-007. Office of
Science and Technology, U.S. Environmental Protection Agency,
Washington, DC. November 2004 [online]. Available: http://www.epa.gov/
waterscience/cs/report/2004/nsqs2ed-complete.pdf [accessed Dec. 26, 2006].

EPA (U.S. Environmental Protection Agency). 2005. Contaminated Sediment
Remediation Guidance for Hazardous Waste Sites. EPA-540-R-05-012.
OSWER 9355.0-85. Office of Solid Waste and Emergency Response, U.S.
Environmental Protection Agency. December 2005 [online]. Available:
http://www.epa.gov/superfund/resources/sediment/pdfs/guidance.pdf [ac-
cessed Dec. 26, 2006].

Johanson, E.E., and J.C. Johnson. 1976. Identifying and Prioritizing Locations for
the Removal of In-Place Pollutants. Contract No. 68-01-2920. Office of Water Planning and Standards, U.S. Environmental Protection Agency, Washington, DC. May 1976.

NRC (National Research Council). 1989. Contaminated Marine Sediments: Assessment and Remediation. Washington, DC: National Academy Press.

NRC (National Research Council). 1997. Contaminated Sediments in Ports and
Waterways. Washington, DC: National Academy Press.

NRC (National Research Council). 2001. A Risk-Management Strategy for PCB-
Contaminated Sediments. Washington, DC: National Academy Press.

Wenning, R.J., M. Sorensen, and V.S. Magar. 2006. Importance of implementation
and residual risk analyses in sediment remediation. Integr. Environ. Assess. Manage. 2(1):59-65.

Zeller, C., B. Cushing. 2006. Panel discussion: Remedy effectiveness: What works,
what doesn't? Integr. Environ. Assess. Manage. 2(1):75-79.

2

Sediment Management at Superfund Megasites

A variety of subjects including environmental engineering, toxicology, environmental monitoring, human and environmental risk assessment, and risk management are relevant to evaluating remediation at contaminated sediment Superfund sites. In this chapter, a number of issues are briefly introduced to provide background for later discussions. Topics include the Superfund process and information available on contaminated sediment Superfund sites; evaluating and managing risks posed by contaminated sediments; and techniques for managing and remediating contaminated sediment with a focus on dredging technologies and their performance capabilities and limitations. The chapter is intended to provide a cursory overview of the topics while emphasizing other sources containing more detailed discussions.

OVERVIEW OF SUPERFUND AND SEDIMENT MEGASITES

Superfund and Environmental Remediation

In 1980, Congress enacted the Comprehensive Environmental Response, Compensation, and Liability Act (CERCLA, 42 U.S.C. 9601-9675), which authorized the establishment of the Superfund program. The goal of the program is to reduce current and future risks to human health and the environment at sites contaminated with hazardous substances. CERCLA established a wide-ranging liability system that makes those responsible for the contamination at sites liable for cleanup costs (see Probst et al. 1995 for greater detail). It also created the "Superfund," a trust fund stocked primarily by a dedicated tax on oil and chemical companies, to fund cleanup activities where there was no financially viable responsible party. Since the taxing authority expired in 1995, the trust fund is largely depleted, and Congress now funds the program from general revenues through annual appropriations (Fletcher et al. 2006).[1] The U.S. Environmental Protection Agency (EPA) implements the program through the National Oil and Hazardous Substances Pollution Contingency Plan (40 CFR § 300), commonly referred to as the NCP or the national contingency plan.

Most of the Superfund program's efforts are aimed at cleaning up sites on the National Priorities List (NPL). Typically, a site is proposed for inclusion on the NPL after being evaluated with a hazard-ranking system, which assesses the potential for hazardous-substance releases at a site to harm human health or the environment (40 CFR § 300 Appendix A). The Superfund process progresses from an initial site assessment through cleanup and eventually deletion of the site from the NPL. Site activities can be paid for by EPA (known as "fund-led" cleanups),[2] by parties connected to the site (referred to as responsible parties), or by some combination of the two.

Selection of a remedy begins with a remedial investigation and feasibility study (RI/FS). The RI is intended to determine the nature and extent of contamination and estimate the associated risk to people and the environment. The FS analyzes and compares remedial alternatives according to the nine NCP criteria (Box 2-1). The criteria require that the remedy, above all, be protective of human health and the environment and comply with all applicable or relevant and appropriate requirements

---

[1]It is worth noting that, in the last few years, EPA has been in the position of not having enough funds to fund all the new remedies that are ready to be started at NPL sites (EPA 2004a).

[2]For fund-lead cleanups, states are required to pay 10% of the costs.

BOX 2-1 Evaluation Criteria for Superfund Remedial Alternatives

Before a remedial strategy is selected for a Superfund site, the options are evaluated on the basis of nine criteria (see below). The first two, overall protection of human health and the environment and compliance with applicable or relevant and appropriate requirements (ARARs), are termed threshold criteria, and a potential remedy must meet them to be selected as a final remedy.[3] The next five criteria are termed balancing criteria and are used in weighing the advantages and disadvantages of potential remedies. The final two criteria are modifying criteria, and the agency is supposed to take them into consideration as part of the selection process.

Threshold Criteria

- *Overall protection of human health and the environment.* This criterion is used to evaluate how the alternative as a whole achieves and maintains protection of human health and the environment.
- *Compliance with applicable or relevant and appropriate requirements (ARARs).* This criterion is used to evaluate whether the alternative complies with chemical-specific, action-specific, and location-specific ARARs or a waiver is justified.

Balancing Criteria

- *Long-term effectiveness and permanence.* This criterion includes an evaluation of the magnitude of human health and ecologic risk posed by untreated contaminated materials or treatment residuals remaining after remedial action has been concluded (known as residual risk) and of the adequacy and reliability of controls to manage such risk. It also includes an assessment of the potential need to replace technical components of the alternative.
- *Reduction of toxicity, mobility, and volume through treatment.* This criterion refers to the evaluation of whether treatment processes can be used, the amount of hazardous material treated (including the principal threat that can be addressed), the degree of expected reductions, the degree to which the treatment is irreversible, and the type and quantity of treatment residuals.
- *Short-term effectiveness.* This criterion includes an evaluation of the effects of the alternative during the construction and implementation phase until

(Continued on next page)

---

[3]Except that specific ARARs can be waived.

BOX 2-1 Continued

remedial objectives are met. It includes an evaluation of protection of the community and workers during the remedial action, the environmental effects of implementing the remedial action, and the expected length of time until remedial objectives are achieved.

• *Implementability.* This criterion is used to evaluate the technical feasibility of the alternative—including construction and operation, reliability, and monitoring—and the ease of undertaking an additional remedial action if the remedy fails. It also considers the administrative feasibility of activities needed to coordinate with other offices and agencies—such as for obtaining permits for off-site actions, rights of way, and institutional controls—and the availability of services and materials necessary for the alternative, such as treatment, storage, and disposal facilities.

• *Cost.* This criterion includes an evaluation of direct and indirect capital costs, including costs of treatment and disposal; annual costs of operation, maintenance, and monitoring of the alternative, and the total present worth of these costs.

Modifying Criteria

• *State (or support agency) acceptance.* This criterion is used to evaluate the technical and administrative concerns of the state (or the support agency, in the case of state-lead sites) regarding the alternatives, including an assessment of the state's or support agency's position and key concerns regarding the alternative, and comments on ARARs or the proposed use of waivers. Tribal acceptance is also evaluated under this criterion.

• *Community acceptance.* This criterion includes an evaluation of the concerns of the public regarding the alternatives. It determines which component of the alternatives interested persons in the community support, have reservations about, or oppose.

Source: Adapted from EPA 2005a.

(ARARs).[4] Remedies are also compared on whether they are technically feasible and cost-effective, provide long-term (permanent) effectiveness, and minimize deleterious effects and health risks during implementa-

---

[4]ARARs pertain to federal, state, or tribal environmental laws relevant to a site.

tion. There is a preference for remedies that can reduce the toxicity, mobility, and volume of contaminants. Finally, there is a preference for remedies that have state and community support.

EPA uses the FS to identify each alternative's strengths and weaknesses and the trade-offs that must be balanced for the site in question (EPA 1988). The agency then selects a remedy and describes it in a record of decision (ROD). Additional studies may be conducted to support the design of the remedy. Once constructed and implemented, the remedy is maintained and monitored to ensure that it achieves its long-term goals. EPA may delete a site from the NPL when a remedy has been implemented, the cleanup goals have been achieved, and the site is deemed protective of human health and the environment (EPA 2000).

If, after implementation of a remedy, contamination exists that could limit potential uses of the site, the site is subject to 5-year reviews even if it has been deleted from the NPL (EPA 2001). The reviews are intended to evaluate the performance of the remedy in protecting human health and the environment and are to be based on site-specific data and observations. However, monitoring is not limited to sites where 5-year reviews are required. EPA guidance states that "most sites where contaminated sediment has been removed also should be monitored for some period to ensure that cleanup levels and RAOs [remedial action objectives] are met and will continue to be met" (EPA 2005a, p. 2-17). Post-remediation monitoring (required in conjunction with 5-year reviews or otherwise) is the basis for evaluating remedy effectiveness and adapting remedial strategies and risk management to achieve remedial action objectives (for further discussion, see Chapter 5).

Sediment Contamination at Superfund Sites

Contaminated sediment is a widespread problem in the United States (EPA 1994, 1997, 1998, 2004a, 2005a). Its wide distribution results from the propensity of many contaminants discharged to surface waters to accumulate in sediment or in suspended solids that later settle. Contaminants can persist in sediment over long periods if they do not degrade (for example, metals) or if they degrade very slowly (for example, polychlorinated biphenyls [PCBs] or polycyclic aromatic hydrocarbons

[PAHs]). Historically contaminated sediment can become buried or, if it is resuspended, can settle out eventually and lie on the sediment surface.

At the national level, the geographic extent of areas with contaminated sediment is not fully defined. In the 2004 Contaminated Sediment Report to Congress (EPA 2004b), EPA reported on sediment sampling at 19,398 sampling stations nationwide, located in about 9% of the waterbody segments in the United States. Of that nonrandom sample of sediment sampling stations, EPA classified 43% as having probable adverse effects, 30% having possible adverse effects, and 27% as having no indications of adverse effects. The 2005 *Contaminated Sediment Remediation Guidance for Hazardous Waste Sites* (EPA 2005a) cites EPA fish advisories covering all five Great Lakes, 35% of the nation's other lakes, and 24% of total river miles as due partly to sediment contamination (EPA 2005b).

EPA does not maintain a current list of NPL sites with contaminated sediments, nor does it compile a list of contaminated sediment areas that are potential Superfund sites. It also does not maintain a list of contaminated sediment sites that are being (or have been) remediated under another authority. EPA did report that "as of September 2005, Superfund has selected a remedy at over 150 sediment sites" (EPA 2006a). In addition, the EPA Office of Superfund Remediation and Technology Innovation is tracking progress at 66 sites, termed tier 1 sites, where the sediment-cleanup remedy involves more than 10,000 cubic yards (cy) of sediment to be dredged or excavated or more than 5 acres to be capped or monitored for natural recovery (EPA 2006b).[5] Of the aforementioned 150 NPL sites where remedies have been selected, EPA considers 11 to be sediment megasites, defined as sites where the sediment portion of the remedy is expected to cost $50 million or more.[6] Of these 11 sites, 10 were proposed for inclusion on the NPL in the very early years of the Superfund program (in 1982-1985), and one (Onondaga Lake) was proposed for inclusion in 1993. Thus, the overwhelming ma-

---

[5]The exact number of tier 1 sites is not clear. EPA's website (EPA 2006b) lists 66 sites while 60 sites are listed in output from EPA's internal database of tier 1 sites (EPA, unpublished data, "Remedial Action Objectives for Tier 1 Sites," Sept. 5, 2006). Seven sites listed in EPA's internal database are not on the website; 13 sites listed on the website are not in the September 5 submittal.

[6]Typically, megasites are defined as sites where the total cost of the remedy *for the entire site* (not just the sediment portion) is expected to be at least $50 million.

jority of the megasites have been on the NPL for over 20 years. Only one of the 11, Marathon Battery, has been formally deleted from the NPL. In addition to the 11 megasites on the NPL, EPA lists two megasites that have been proposed for the NPL but are not final (GE Housatonic River, MA and Fox River, WI) and one that has not been proposed (Manistique River/Harbor area, MI). The 14 sites are listed in Table 2-1. The status of remediation at the sites varies. At some, such as Bayou Bonfouca and Marathon Battery, remediation has been completed; at others, such as Commencement Bay and Sheboygan Harbor, remedial activities are going on; and at still others, such as Hudson River and Onondaga Lake, remedial activities have not begun. Megasites are described only in terms of remediation cost (at least $50 million), so the size and volume of contaminated materials at the sites can vary greatly (see Box 2-2).

One might ask, Why all this attention to contaminated sediment megasites if there are only 14 nationwide? There are two reasons. First, at 13 of the megasites mentioned above (no cost information was provided on the Triana/Tennessee River site), total remedial costs are estimated to be about $3 billion, a huge amount of money even by Superfund standards.[7] Second, the 14 sites probably constitute only a subset of the contaminated sediment sites that will entail expensive remedies and will be cleaned up under the Superfund program. For example, the EPA list of contaminated sediment megasites does not include some well-known sites, such as the Bunker Hill Mining and Metallurgical Complex, ID, and Love Canal, NY. Both those tier 1 sites are megasites by the conventional definition (total remediation cost of at least $50 million), but the sediment portion alone is not expected to be $50 million.

When comparing EPA's list of tier 1 sites (EPA 2006b) with a somewhat dated list of megasites[8] (that does not include federal facilities), one can find 11 "conventional" megasites on the tier 1 list.

- Alcoa–Point Comfort/Lavaca Bay, TX
- Allied Paper Inc./Portage Creek/Kalamazoo River, MI
- Bunker Hill Mining and Metallurgical Complex, ID

---

[7]Based on data provided by EPA, "50M cost Query_091306.xls" (EPA, unpublished data, Sept. 18, 2006).

[8]Based on the report to Congress, *Superfund's Future: What Will It Cost?* (Probst et al. 2001), which lists megasites as of FY 2000.

TABLE 2-1 Sediment Megasites (Sites at Which Remediation of the Sediment Component Is Expected To Be at Least $50 million)

Site Name, State
NPL Sites
New Bedford Harbor, MA
Hudson River PCBs, NY
Marathon Battery Corp., NY
Onondaga Lake, NY
Triana/Tennessee River, AL
Sheboygan Harbor and River, WI
Velsicol Chemical, MI
Bayou Bonfouca, LA
Milltown Reservoir Sediments, MT
Silver Bow Creek/Butte Area, MT
Commencement Bay, WA
Non-NPL Sites
GE Housatonic River, MA
Fox River, WI
Manistique River/Harbor area, MI

Source: EPA, unpublished data, "$50M Cost query_091306.xls," Sept. 18, 2006.

- Eagle Mine, CO
- EI duPont–Newport landfill, DE
- GM–Central Foundry Division (Massena), NY
- Lipari landfill, NJ
- Love Canal, NY
- McCormick and Baxter Creosoting Co., CA
- Nyanza chemical waste dump, MA
- Wyckoff Co.–Eagle Harbor, WA

Furthermore, as described below, large and expensive sediment remediations are conducted under authorities other than Superfund.

A crucial question is how many additional major contaminated sediment sites are likely to be listed on the NPL. EPA does not designate "likely future megasites" in its tier 1 list of sites or NPL sites for which RODs have not been issued. According to EPA, the most likely future sediment megasites are the "tier 2" contaminated sediment sites (S. Ells,

---

BOX 2-2 How Large Is a Megasite?

Contaminated sediment megasites are among the most challenging and expensive sites on the NPL. Megasites are conventionally defined as those with remedial activities costing at least $50 million, but there are large differences in the magnitudes and scales of these sites. A few megasites, such as Bayou Bonfouca and Marathon Battery, are relatively small, with dredging activities covering tens of acres and operations occurring over a few years. Other dredging projects—such as those in the Fox River, New Bedford Harbor, and Commencement Bay—are components of broader activities at large-scale megasites where remedial activities are going on and will take years or decades to complete. The $50-million distinction for a megasite is not readily translatable into volume of materials removed. For example, sediment remediation (including design, mobilization, marine demolition, dredging, water management, transportation and disposal, construction oversight and EPA oversight, without the upland-based removal costs) at the Head of Hylebos Waterway in Commencement Bay, WA, removed 404,000 cy at a cost of $58.8 million (about $145/cy) (P. Fuglevand, personal commun., Dalton, Olmsted & Fuglevand, Inc., May 11, 2007). In Manistique Harbor, MI, dredging operations removed 187,000 cy at a cost of $48.2 million (about $260/cy) (Weston 2002). Dredging operations in Bayou Bonfouca, LA, removed 170,000 cy at a cost of $90 million (about $530/cy) (EPA, unpublished information, "$50M Cost query_091306.xls," Sept. 18, 2006).

EPA, personal commun., Oct. 12, 2006). Tier 2 sites are designated for review by the Contaminated Sediments Technical Advisory Group because they are large, complex, or controversial contaminated sediment Superfund sites.[9] There are 12 tier 2 sites. Three are on the earlier two lists provided, but nine are not. Of the nine, four are NPL sites (Ashland/Northern States Power, WI, Portland Harbor, OR, Lower Duwamish Waterway, WA, and the Pearl Harbor Naval Complex, HI), and five are not (Palos Verdes, CA, Kanawah River/Nitro, WV, Centredale Manor Restoration Sites, RI, Anniston PCB site, AL, and Upper Columbia River, WA). EPA also indicates that the Passaic River, NJ, Berry's Creek at Ventron/Velsicol, NJ, and Tar Creek, OK, are likely future con-

---

[9]Although, it should be noted that EPA indicates that "No quantifiable criteria were used to develop this list." The list of sites is available at http://www.epa. gov/superfund/resources/sediment/cstag_sites.htm.

taminated sediment megasites, although they have not been designated as tier 2 sites (S. Ells, EPA, personal commun., Sept. 18, 2006).

Because predicting future NPL listings is more an art than a science, in some ways, it is not surprising that there is no official list of likely future contaminated sediment megasites. That said, the committee was surprised that there is so little effort devoted to tracking and understanding likely future sediment megasites at the national level. Apparently, fewer than two full-time employees are assigned to contaminated sediment issues at Superfund headquarters. It appears that EPA has not allocated the resources needed to identify the scope of the problem and to develop a strategy to address issues related to contaminated sediments. To develop an effective long-term contaminated sediment strategy it is critical to know how much work remains to be done. To address that question, one needs to have three pieces of information:

1.   How much work remains at sites already categorized as contaminated sediment megasites.

2.   How many contaminated sediment sites already on the NPL are likely to be determined to be megasites.

3.   How many new such sites are likely to be added to the NPL in the coming years.

None of that information is readily available from EPA. Clearly, EPA should not stop and wait until this information is collected. However, it is important that EPA obtain this information and update it regularly in order to be able to forecast likely future costs and needed resources, as well as to assess what kinds of research and monitoring improvements are likely to have the largest benefit to the program.

Cleanup Under Authorities Other Than Superfund

Remediation of contaminated sediments is also conducted under authorities other than Superfund and can be led by various parties, such as state or federal agencies or private entities, in combination or individually. For example, a 5-mile reach of the Grand Calumet River, a highly industrialized tributary to Lake Michigan in northwest Indiana, was dredged by U.S. Steel Corporation pursuant to a Clean Water Act consent decree and a Resource Conservation and Recovery Act corrective-action consent order (Menozzi et al. 2003). This project, described as "the largest environmental dredging project to be undertaken in North America," removed 786,000 cy of sediment from the Grand Calumet River (U.S. Steel 2004).

State programs conduct and oversee sediment remediation under a variety of authorities. For example, the State of Washington Department of Ecology is charged with cleaning up and restoring contaminated sites under authority of the Model Toxics Control Act (MTCA) and Sediment Management Standards (SMS) (Washington Department of Ecology 2005). In 2005, 142 sediment cleanup sites were identified in Washington: 41 were being cleaned up under federal authorities, 48 were using state authority alone, 11 were under federal and state authorities, and the remaining 42 were either voluntary (conducted by the responsible party) or the authority had not been assigned (Washington Department of Ecology 2005).

Contaminated sediments in many harbors and rivers of the Great Lakes are addressed in the Great Lakes Water Quality Agreement between the United States and Canada, which established 43 areas of concern (AOCs) in U.S. and Canadian waters. The U.S. EPA Great Lakes National Program Office administers funds from the Great Lakes Legacy Act of 2002 for the remediation of contaminated sediment at AOCs (EPA 2004c). The first Legacy Act cleanup was in 2005 at the Black Lagoon in the Detroit River AOC near Trenton, MI. At that site, 115,000 cy of contaminated material was dredged, and the area was capped. Hog Island, near Superior, WI, in the St. Louis River AOC of Lake Superior, was remediated with dry excavation (see Sediment Management Techniques in this chapter for a description of remedial methods). In 2006, two projects were under way with Great Lake Legacy Act funds. The Ruddiman Creek remedial action in Muskegon, MI, contains an excavation and dredging component and is expected to remove around 80,000 cy. Dredging will also occur at the Ashtabula River, near Cleveland, OH, where it is expected that about 600,000 cy of contaminated sediment will be removed from the lower portion of the river.

Another program, the Urban Rivers Restoration Initiative, is a collaboration between EPA and the U.S. Army Corps of Engineers for urban-river cleanup and restoration (EPA 2003a). Eight demonstration pilot projects, including a dredging project in the Passaic River in New

Jersey, have been developed to coordinate the planning and implementation of projects to promote clean water and sediment among multiple jurisdictions and federal authorities.

This section is by no means a comprehensive listing of sediment projects or efforts outside of Superfund; rather, the intent is to convey that there are many sediment remediation projects outside of Superfund that are conducted by multiple groups and under several authorities. To the extent that other environmental dredging activities are conducted to address risk from contaminated sediments, many of the discussions and conclusions presented in the latter chapters of this report will be applicable.

EVALUATING RISK REDUCTION AT CONTAMINATED SEDIMENT SITES

Risks Posed by Contaminated Sediment

As briefly described in Chapter 1, contaminants in sediment can pose risks to human health and the environment. Apart from direct exposure to contaminated sediment during, for example, recreational activities, humans typically are exposed to contaminants through the ingestion of fish or wildlife that have accumulated contaminants from the sediment. Fish and wildlife are exposed to contaminants in sediments through a number of pathways, including absorption from pore water or sediments, incidental ingestion of contaminated sediments, and consumption of contaminated organisms. Several of those processes are presented graphically in Figure 2-1. Predicting effects of exposures can be complex. Variations in the sediment environments will alter the bioavailability of contaminants, and this can markedly affect their accumulation and effects on organisms (NRC 2003). For instance, the presence of sulfide will greatly decrease the bioavailability of many metals, and organic carbon can decrease the bioavailability of organic pollutants, such as PCBs (EPA 2003b, 2005c).

Accessibility is a primary factor in exposure to and effects of contaminated sediments. A common problem in assessing risks posed by contaminated sediments is that the contaminants (or the highest contam-

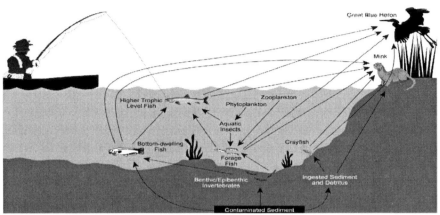

FIGURE 2-1 Generic conceptual site model indicating contaminated sediment exposure pathways between sediment and ecologic receptors, including fish, shellfish, benthic invertebrates, birds, and mammals. Source: EPA 2005a.

inant concentrations) can be buried beneath relatively clean sediment that has deposited over time (see Figure 2-2).

Because sediment contaminants typically are strongly associated with the sediment particles, contaminants buried below the biologically active zone are neither accessible nor available to sediment- or water-dwelling organisms. In such cases, a relatively small continuing source may pose a greater risk of exposure and associated injury than a large buried inventory of sediment associated contaminants. Risk due to sediments is usually limited to contaminants that are present in or can migrate into the biologically active zone, the upper layers of sediment where organisms live or interact. That layer typically ranges from a few centimeters to 10-15 cm deep, although some organisms (including aquatic plants) may penetrate more deeply (Thoms et al. 1995; NRC 2001).

Sediment-associated contaminants tend to collect in relatively stable depositional zones in water bodies. In such environments, buried contaminants (that is, those below 10-15 cm) may never be exposed to the biologically active zone. However, water bodies are dynamic systems and even in generally depositional and stable environments, high flow events and changes in hydrologic conditions may lead to short term

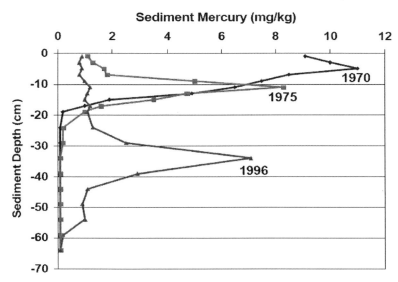

FIGURE 2-2 Historical changes in sediment core profiles of mercury concentrations collected from Bellingham Bay, WA, during natural recovery (dredging or capping was not performed). Sediment cores were taken over time (1970, 1975, 1996) from the same vicinity of Bellingham Bay in an area with a stable sediment deposit. At this site, mercury was released from a nearby facility from 1965 until controls were put in place in 1971. Sedimentation has since continued to bury the contaminants. Source: Patmont et al. 2004. Reprinted with permission from the authors; copyright 2003, Battelle Press.

erosion, exposure, and transport of these contaminants to the biologic active zone. In environments subject to such conditions, removal of contaminant mass may be an effective remedial response to the risk posed by them. If contaminants buried below the biologically active zone are likely to remain buried, the potential exposure and risk may be so small that remediation of any kind is unwarranted. Remediation of deeply buried contaminated sediments that do not contribute to the exposure of aquatic systems now or under future conditions will not achieve risk reduction goals. In such cases, other contaminant sources, for example inadequately controlled surface discharges or atmospheric deposition, may control exposure and risk. A fair amount of effort in recent years has gone into developing approaches for assessing sediment column stability and refining hydrodynamic models and linking them with fate and transport models to estimate contaminant transport under various conditions (e.g., Bohlen and Erickson 2006). Output from these approaches and models are important in estimating the risks associated with remedial alternatives.

Decision-Making in a Risk-Based Framework

Principles for understanding and comparing risk reduction from various sediment remediation techniques are discussed briefly below, however, it should be noted that it is not the mandate or intent of the report to develop specific recommendations and procedures for performing comparative risk analyses of remedial alternatives in selection of a sediment remedy. While important, that type of detailed assessment was not requested or undertaken. The brief discussion provided here on risk-based remedy selection is intended to provide background for later discussions on improving decision making.

The process of managing risk at contaminated sediment sites was evaluated extensively in the 2001 National Research Council report *A Risk Management Strategy for PCB-Contaminated Sediments* (NRC 2001). Perhaps its most relevant conclusion is that all decisions regarding the management of PCB-contaminated sediments should be made in a risk-based framework. The report further suggests that the framework developed by the Presidential/Congressional Commission on Risk Assessment and Risk Management provides a good foundation for assessing the risks and the management options for a site (see Box 2-3). The general framework exhibits several key features that make it appropriate for the management of contaminated sediment sites. It recognizes that risk-reduction should be the foundation of any decision-making process and the importance of the participation of interested and affected stakeholders in the decision-making process. It also provides a systematic and structured process for identifying and assessing risks, evaluating and implementing options, and monitoring the success of the overall process. The 2001 Research Council report also recommends that risk assessments and risk management decisions be site specific and concluded that current management options can reduce risks but cannot eliminate PCBs and PCB exposure from contaminated sediment sites. Because all remedial options will leave residual PCBs, the short- and long-term risks that they pose should be considered in evaluating management strategies. Those ideas also apply to other sediment contaminants.

BOX 2-3 Presidential/Congressional Commission on Risk Assessment and Risk Management

The Presidential/Congressional Commission on Risk Assessment and Risk Management was formed in response to the 1990 Clean Air Act Amendments in which Congress mandated that a Commission on Risk Assessment and Risk Management be formed to "make a full investigation of the policy implications and appropriate uses of risk assessment and risk management in regulatory programs under various Federal laws to prevent cancer and other chronic human health effects which may result from exposure to hazardous substances" (PCCRARM 1997, p. i).

The commission ultimately developed a report that introduced a risk management framework "to guide investments of valuable public sector and private sector resources in researching, assessing, characterizing, and reducing risk" (PCCRARM 1997, p. i). The commission proposed a six-stage process:

- Define the problem and establish risk management goals.
- Assess risks associated with the problem.
- Evaluate remediation options for addressing the risks.
- Select a risk management strategy.
- Implement the risk management strategy.
- Evaluate the success of the risk management strategy.

This process should be conducted

- In collaboration with all affected parties.
- In an iterative fashion when substantive new information becomes available.

The proposed process, depicted in the schematic below, is a systematic method to manage risks that the commission defined as "the process of identifying, evaluating, selecting, and implementing actions to reduce risk to human health and to ecosystems. The goal of risk management is scientifically sound, cost-effective, integrated actions that reduce or prevent risks while taking into account social, cultural, ethical, political, and legal considerations" (PCCRARM 1997, p. 2).

Key features of the framework are recognition of the importance of stakeholders in the process, the importance of defining risks in a broader context than single risks associated with single chemicals in single environmental media, and the importance of an iterative process where earlier decisions can be revisited when new findings are made.

Remedy selection is a complex process with many considerations (see Box 2-1). In some cases, removal will be the best option for risk reduction and satisfying the NCP criteria, in others, capping or monitored natural recovery will be preferable. An analysis of alternative remedies typically includes a comparison of both the short- and long-term risks to human and environmental receptors associated with a particular site. For example, the risks from dredging can include exposure to contaminants during dredging, rehandling, and transport, and contaminants that remain after operations are completed. Those risks would be compared to other alternatives, including risks from unconfined contaminated sediment and potential future resuspension and transport during storm and non-storm events. Risks beyond those related directly to exposure to contaminants are also considered in this process (see Net Risk Reduction below).

Technical and policy guidance for making remedial decisions using a risk based framework at contaminated sediment sites was recently issued (EPA 2005a). The document provides a useful evaluation of the various sediment management approaches and their advantages and limitations in attaining risk reduction. It discusses in detail aspects of the

Superfund decision-making process (site characterization, feasibility study, and remedy selection) particular to contaminated sediment, and it offers recommendations for implementing an effective monitoring plan. The guidance concludes that "The focus of remedy selection should be on selecting the alternative best representing the overall risk reduction strategy for the site according to the NCP nine remedy selection criteria.... EPA's policy has been and continues to be that there is no presumptive remedy for any contaminated sediment site, regardless of the contaminant or level of risk" (EPA 2005a, p. 7-16).

Measuring Risk-Reduction

Estimating the degree of risk reduction is central in considering the potential effectiveness of a remedial action. Risk posed by chemical contamination is a function of the duration and intensity of exposure and the ability of the chemical or chemical mixture to exert adverse effects. There is not a direct measure of risk, so surrogate metrics are used to estimate risk. Environmental analyses have to use metrics that, in practice, can be employed relatively easily, are not time- and cost-prohibitive, and have sufficient accuracy and precision to be reliable. Estimates of risk reduction at contaminated sites have often centered on measuring changes in the mass, volume, and concentration of contaminated sediments. Those measures are related to the potential for exposure in the aquatic environment but do not provide information on effects. Therefore, although they are the most prevalent, they are not fully adequate to describe risk or to chart risk reduction. Toxicity testing, biologic community indexes, and tissue-residue analyses provide a fuller picture of effects, although they too have limitations in their ability to describe risk. (See Chapter 5 for further discussion on the metrics and their advantages and disadvantages in estimating risk to aquatic biota and humans.)

Temporal Scale

In characterizing risk and evaluating risk reduction, it is necessary to consider the duration of time over which exposure and effects occur. After remediation, risk is usually predicted to decline over time rather than reach a protective level immediately on completion of the remedy, so remedy selection involves a comparison of time profiles of predicted risks, often on scales of decades. Such a comparison of risk profiles over time is how the long-term effectiveness of dredging is evaluated relative to alternative technologies for the largest and most complex sites. Factors dictating the time to reduce risk are site specific and include the time required to design and fully implement the remedy, the time required to cleanse the food chain of existing contaminant body burdens, and the time for natural recovery processes to attenuate any residual surface sediment concentrations after implementation is complete.

Net Risk Reduction

The 2001 National Research Council report indicated that the paramount consideration for contaminated sites should be the management of overall or net risks to humans and the environment in addition to specific risks. The report concludes that the evaluation of sediment management and remediation options should take into account all costs and potential changes in risks for the entire sequence of activities and technologies that constitute each management option. (For example, managing risks from contaminated sediments in aquatic environments might result in the creation of additional risks in both aquatic and terrestrial environments.) The report also suggests that a broader array of risks—including societal, cultural, and economic risks—should be evaluated comprehensively. The concept of net risk reduction has been embraced by EPA in its *Contaminated Sediment Remediation Guidance for Hazardous Waste Sites* (EPA 2005a); it states that "Project managers are encouraged to use the concept of comparing net risk reduction between alternatives as part of their decision-making process for contaminated sediment sites, within the overall framework of the NCP remedy selection criteria. Consideration should be given not only to risk reduction associated with reduced human and ecologic exposure to contaminants, but also to risks introduced by implementing the alternatives. The magnitude of implementation risks associated with each alternative generally is extremely site specific, as is the time frame over which these risks may apply to the site. Evaluation of both implementation risk and residual risk are existing important parts of the NCP remedy selection process" (EPA 2005a, pp. 7-13, 7-14).

Risk-Based Objectives and Cleanup Levels

Each site has its own set of contaminants with different concentrations and distributions in its own particular geologic, geochemical, geographic, social, ecologic, and economic setting. Therefore, management decisions based on the above framework are expected to differ among sites.

At Superfund sites, the overall goal of sediment management is reduction of risk to human health and the environment. That goal takes the form of remedial action objectives, which are used in developing and comparing alternatives for a site, and typically describe the desired effect of the remediation on risk (for example, reduction to acceptable levels of the risks to people ingesting contaminated fish). Attainment of remedial action objectives can be difficult to quantify or might occur in a time frame or encompass a spatial scale that makes it difficult to link to remedial actions. Under such circumstances, cleanup levels, such as achievement of a sediment concentration or removal of a given mass of contaminant or sediment, which can be more easily used to evaluate remedial actions, are often adopted (EPA 2005a). Ideally, clean-up levels are tied to effects-based risk thresholds, and take into account effects of combinations of contaminants.

That the application of a good risk management strategy is likely to result in significantly different cleanup levels at different sites makes it difficult for the committee to draw conclusions about the expected effectiveness of dredging. At some sites, cleanup levels are far less stringent than at others, and thus all other things being equal, a site with less stringent cleanup goals is more likely to be "successful" than a site with more stringent goals. Geologic and site-specific conditions also differ, so even if the cleanup goals are similar at different sites, the technical ability to reach the goals may differ. Thus, one needs to be highly cautious in suggesting that success with a remedial option at one site necessarily means that the same success is likely at another.

To help to ensure that remedial actions achieve their desired objectives, 11 principles have been developed by EPA to guide sediment remediation (Box 2-4). The principles were developed partially in response to the recommendations of the National Research Council Committee on Remediation of PCB-Contaminated Sediments (NRC 2001).

BOX 2-4 Eleven Principles of Contaminated Sediment Management

In February 2002, the EPA Office of Solid Waste and Emergency Response promulgated 11 principles of contaminated sediment management (OSWER Directive 9285.6-08):

1.   Control sources early.
2.   Involve the community early and often.
3.   Coordinate with states, local governments, tribes, and natural-resources trustees.
4.   Develop and refine a conceptual site model that considers sediment stability.
5.   Use an iterative approach in a risk-based framework.
6.   Carefully evaluate the assumptions and uncertainties associated with site-characterization data and site models.
7.   Select site-specific, project-specific, and sediment-specific risk-management approaches that will achieve risk-based goals.
8.   Ensure that sediment cleanup levels are clearly tied to risk-management goals.
9.   Maximize the effectiveness of institutional controls and recognize their limitations.
10. Design remedies to minimize short-term risks while achieving long-term protection.
11. Monitor during and after sediment remediation to assess and document remedy effectiveness.

The principles were designed to help EPA site managers to make scientifically sound and nationally consistent risk-management decisions at contaminated sediment sites. The principles are consistent with the recommendations of the National Research Council *A Risk Management Strategy for PCB-Contaminated Sediments* and the Presidential/Congressional Commission on Risk Assessment and Risk Management. They were incorporated into the *Contaminated Sediment Remediation Guidance for Hazardous Waste Sites* (EPA 2005a).

The Conceptual Site Model: A Working Understanding of Processes Leading to Risk from Sediment Contamination

The development of remedial action objectives and cleanup levels to reduce risk is based on a conceptual understanding of cause-effect relationships among contaminant sources, transport mechanisms, exposure pathways, human receptors, and ecologic receptors at each affected level of the food chain. That understanding of causal relationships is known as a conceptual site model (CSM) (EPA 2005a) and is typically developed for each site on the basis of site-specific conditions. The link between risk and the inventory of contaminants in sediments is not always obvious. An accurate CSM is critical for identifying the processes and pathways that might lead to risk and appropriate means of intervening to reduce risk. For example, evaluation of the stability or potential instability of buried deposits and their potential for exposure in the bioavailable zone is a key component of a CSM because a CSM must be able to differentiate between important and unimportant routes of exposure.

In addition to linking site contaminant sources to exposures and risks, the CSM must account for background conditions, including contaminant distribution from offsite sources. Ecosystems may be highly stressed because of multiple watershed and atmospheric effects on conventional water quality measures, such as nutrients, suspended solids, acidity, dissolved oxygen, and temperature. The contribution of background stressors or other background sources to site effects should be evaluated, including assessment of their importance relative to the contaminants of concern, and recognized as potentially complicating factors in ecosystem restoration.

The CSM should guide site investigation, and its hypotheses and assumptions should be tested and refined as site data are acquired. When the CSM has been accepted with a high degree of confidence, it is used to define remedial action objectives. Basing remedial action objectives on the best scientific understanding of the mechanisms that lead to site-specific risk maximizes the likelihood that remedial actions will meet the objectives.

For most sites, but especially for the largest and most complex, a quantitative dimension must be added to the CSM to support development and selection of a remedy. The result is a mathematical model or a set of models of the various component processes. Mathematical models are used to quantify the same cause-effect relationships that are embodied in CSMs so that magnitudes of predicted outcomes can be associated with specific causes or actions, such as contaminant loads, environmental conditions, and remedies. To be most accurate, the models should be supported by and calibrated to site-specific data on the environmental media (such as sediments, pore water, and water), receptors (such as benthic organisms, fish, and humans), and processes (such as toxicity, bioavailabity, and bioaccumulation) that are being examined.

Mathematical models range from simple to complex, including analytic equations representing established scientific relationships between independent and dependent variables, statistical cause-effect relationships between site variables, and systems of differential equations representing multiple fate and transport processes. With a mathematical model, quantitative versions of hypotheses can be tested and refined on the basis of site data, including data from field surveys of site conditions and pilot studies of remedial technologies, and then the relative effectiveness of alternative remedies in reducing exposures can be estimated, including the sensitivity of exposure to the remedies and the time needed to reduce exposure. The measures of predicted effectiveness are used to support remedy selection.

Models are subject to uncertainty because of the uncertainty in parameters and process representations. Model testing and refinement does not end with the selection of a remedy through a record of decision. It is important that the conceptual and mathematical site models also be used in designing monitoring of conditions during implementation and post-remedy phases and that the monitoring data be used to validate the models' predictions. When risk reduction deviates significantly from a model's predictions, the model should be modified or recalibrated to improve its accuracy so that more reliable predictions can be available to guide midcourse adjustments in the remedy (EPA 2005a).

CONTAMINATED SEDIMENT MANAGEMENT TECHNIQUES

Contaminated sediment is managed with various techniques, including source control, natural recovery, capping, and removal (dry excavation and dredging). Removal necessitates management of the removed material, which normally includes dewatering, transport, and disposal. Treatment of dredged material to remove or destroy contaminants is an option, but cost and other factors usually lead to disposal in upland landfills or in near-shore confined disposal facilities. In some cases, dredged material can be returned to the aquatic environment through containment in confined aquatic disposal facilities (EPA 2005a).

The National Research Council Committee on Contaminated Marine Sediments (NRC 1997) and Committee on Remediation of PCB-Contaminated Sediments (NRC 2001) have reviewed and reported on a number of sediment management techniques. The committees stated that source control is advisable in all contaminated sediment management projects, notwithstanding the difficulties of identifying some sources of contamination. Beyond source control, interim controls (temporary measures to address exposures immediately) and long-term controls (such as in situ management technologies, sediment removal and transport, and ex situ management) may be needed to address sediment contamination.

More recently, EPA (2005a) lists both in situ and ex situ remedial strategies for managing risks posed by contaminated sediment. The in situ strategies include monitored natural recovery (MNR), in situ capping, hybrid (thin-layer placement) approaches, institutional controls, and in situ treatment; the ex situ strategies include dredging and dry excavation (following dewatering or water diversion). See Box 2-5 for an explanation of these approaches. The present committee's focus is on environmental dredging, which is conducted specifically to remove contaminated sediments, as opposed to navigational dredging, which typically is intended to maintain depth in waterways for navigation or other purposes.

HISTORICAL PERSPECTIVE ON THE USE OF REMOVAL TECHNOLOGIES TO REDUCE RISK

Although there has never been a presumptive remedy for sediments, the historical preference for removal is evident in the large percentage of sites whose remedy was based entirely or in part on dredging. In an overview of Superfund sediment remediation, EPA presented information from 60 tier 1 sites for which a remedy had been selected. Of the 60, 57% had only removal as the remedial action, 15% capping with removal, 13% removal with MNR, 5% only capping, 2% only MNR, and 8% all three remedies (Southerland 2006). The historical preference for

---

BOX 2-5 Remedial Approaches to Contaminated Sediment
In Situ Approaches

In Situ Approaches

• *Monitored natural recovery (MNR)* is a remedy for contaminated sediment that typically relies on naturally occurring processes to contain, destroy, or reduce the bioavailability or toxicity of contaminants in sediment.
• *In situ capping* refers to the placement of a subaqueous covering or cap of clean material over contaminated sediment, which remains in place.
• *Hybrid approaches* refers to placement of a thin layer of sand or other material to accelerate recovery.
• *Institutional controls* are controls on the use of resources. They typically include fish-consumption advisories, commercial fishing bans, waterway-use or land-use restrictions (for example, no-anchor or no-wake zones and limitations on navigational dredging), and agreements on maintenance of dams or other structures.

Ex Situ Approaches

• *Dredging and excavation* are common means of removing contaminated bottom sediment from a body of water either while the sediment is submerged (dredging) or after water has been diverted or drained (excavation). Ex situ approaches can include backfilling with clean material as needed or appropriate.

Source: Adapted from EPA 2005a.

---

removal is probably based on the perception (in both agencies and the public) of the permanence of the remedy. Dredging and excavation remove the mass of contaminants from the aquatic environment, and this has historically been viewed as key to reducing human health and environmental risks.

Technologies for removing sediment were already well established in the early years of sediment cleanup, in part as an extension of remediation technologies applied at upland sites. Most of the initial technologies for managing sediment came from the U.S. Army Corps of Engineers experience with navigational dredging and disposal. Other remedies were typically viewed as less certain by regulators and the public with respect to long-term effectiveness or permanence. Leaving contamination in place under a capping or MNR remedy was often considered more uncertain because of the residual risk posed by contaminants left in place.

The dynamic nature of aquatic environments has often led to the selection of removal as the preferred alternative in many areas of the country. Contaminated sediment is often associated with industrial, urban harbors where operational and navigational constraints are viewed as limiting the feasibility of capping or natural recovery. Those environments are often subject to disturbances, such as those caused by prop wash, seasonal flooding, ice scour, and storm surges, which were viewed as creating substantial risk if contaminants were left in place.

Removal of contaminated sediment has brought unique challenges that were initially not well recognized. Navigational dredging techniques adopted for environmental dredging are designed to achieve a specific bottom elevation or the removal of a specific volume, often in the shortest possible time, whereas environmental dredging typically must achieve a specific final concentration while minimizing contaminant releases during dredging, handling, and disposal. As dredging remedies have been implemented at various sites, the effects of resuspension and transport of contaminated material off site and residual contamination in a remediated area have become apparent (Bridges et al. in press). The risks associated with the implementation of environmental dredging have received a great deal of attention in the last few years (EPA 2005a; Wenning et al. 2006).

Greater experience with capping remedies has been gained over the last decade; cap performance can be better predicted and quantified, and this has led to greater acceptance among agencies. In addition, capping typically has been less expensive and can be implemented more quickly, so it is often preferred by responsible parties (Palermo et al. 1998). In response to the increasing experience with remedial technologies, recent guidance from EPA has called for a more equitable evaluation of all remedies with careful analysis of the short-term and long-term risks associated with any remedy and thorough consideration of site-specific conditions (EPA 2005a).

OVERVIEW OF ENVIRONMENTAL DREDGING

Dredging refers to the removal of sediment from an underwater environment. It involves dislodging and removing material on the bottom of a waterway. Dredges are normally classified according to the basic operation by which sediment is removed, such as mechanical or hydraulic[10] (EPA 1994). For purposes of this report, excavation in the dry using conventional equipment operating within dewatered containments such as sheet-pile enclosures or cofferdams is not covered. The term environmental dredging is more generally associated with removal of sediment from under water. Environmental dredging can be accompanied by backfilling of the dredged areas. Placement of clean material covers and mixes with dredging residuals and further reduces risk from contamination that remains after dredging. Unlike capping, permanent confinement of underlying material is usually not the goal.

Typical objectives of environmental dredging are shown in Box 2-6. Because the purpose of navigational dredging is to restore navigable depth to a waterway, the selection of equipment and operational approaches considers economics, effectiveness, and environmental protection (USACE/EPA 1992) in that order. Conversely, environmental dredging has remediation as its stated purpose. The distinction results in reversing the order of importance of the selection factors for equipment and operational approaches; that is, one needs to consider environmental protection and effectiveness first before considering economics (Palermo et al. 2006).

BOX 2-6 Objectives of Environmental Dredging

- Dredge with sufficient accuracy such that contaminated sediment is removed and cleanup levels are met without unnecessary removal of clean sediment.
- Dredge the sediments in a reasonable period of time and in a condition compatible with subsequent transport for treatment or disposal.
- Minimize and/or control resuspension of contaminated sediments, downstream transport of resuspended sediments, and releases of contaminants of concern to water and air.
- Dredge the sediments such that generation of residual contaminated sediment is minimized or controlled.

Source: Palermo et al. 2006.

---

[10]Pneumatic systems, which use compressed air to pump sediment out of a waterway, have not gained general acceptance in environmental dredging projects in the United States.

Types of Environmental Dredges

Selection of dredging equipment is sediment specific, site specific, and operations specific. Many textbooks and manuals describe the science and engineering principles of dredges, their selection, and their operation (Bray 1979; USACE 1983; Herbich 2000). This section provides basic definitions of dredging methods and equipment types normally considered for environmental dredging. There is no attempt to list all the possible types of dredge equipment that may be applicable to environmental dredging. Box 2-7 lists the equipment most commonly used for environmental dredging according to type (category) and definition (Palermo et al. 2004). Figure 2-3 shows the basic dredge types. More detailed descriptions of environmental-dredging equipment are available elsewhere (Averett et al. 1990; EPA 1994; EPA 2005a).

Other dredge types—such as hopper dredges, dustpan dredges, and bucket-ladder dredges—are not included in Box 2-7, because they are used primarily for navigational dredging. In addition, within dredge types, specific designs may differ and may have varied capability. In general, the dredge types listed above represent equipment that is readily available and used for environmental dredging projects in the United States.

A number of newer dredges, including some specifically designed for environmental dredging, are available. They have been termed specialty dredges and are intended to provide benefits by reducing sediment resuspension and contaminant releases. Other advantages may include operational efficiency for removal of sediment and transportation, depending on the sediment and project conditions and the performance standards. Most specialty dredge designs originated outside the United States, but several U.S. companies have now formed partnerships that allow use of specialty equipment from various countries. Field experience with specialty dredges in the United States is limited (Palermo et al. 2003). The dredges have been proposed for use at contaminated sediment sites, but little information is available about their sediment-extraction efficiency or about the claimed improvements in innovations, such as improved solids capture and reduced resuspension.

The equipment used for environmental dredging is usually smaller than that commonly used for navigation dredging because removal

---

BOX 2-7 Equipment Commonly Used in Environmental Dredging

Mechanical Dredges

- Clamshell: Wire-supported conventional open clam bucket.
- Enclosed bucket: Wire-supported, nearly watertight or sealed bucket. In contrast to conventional open buckets, recent designs also incorporate a level-cut capability instead of the circular cut of conventional buckets (for example, the horizontal profiling buckets).
- Articulated mechanical: Backhoe design, clam-type enclosed bucket, hydraulic closing mechanism, all supported by articulated fixed arm.

Hydraulic Dredges

- Cutterhead: Conventional hydraulic pipeline dredge with conventional cutterhead.
- Horizontal auger: Hydraulic pipeline dredge with horizontal auger dredgehead.
- Plain suction: Hydraulic pipeline dredge using a dredgehead design with no cutting action and plain suction (for example, cutterhead dredge with no cutter basket mounted, matchbox dredgehead, and articulated scoops).

Pneumatic Dredges

- Pneumatic: Air-operated submersible pump, pipeline transport, and wire-supported or fixed-arm-supported.

Specialty Dredges and Diver-Assisted Dredges

- Specialty dredgeheads: Other hydraulic pipeline dredges with specialty dredgeheads or pumping systems.
- Diver-assisted: Hand-held hydraulic suction with pipeline transport.

Source: Adapted from Palermo et al. 2004.

---

volumes and rates tend to be lower and water to be shallower. Mechanical-bucket sizes range from 2 to 8 m^3 (about 3 to 10 cy), and hydraulic-pump sizes range from 15 to 30 cm (about 6 to 12 in.) (Palermo et al. 2006). Obviously, larger dredges are available for both mechanical and hydraulic equipment and can be used for environmental dredging if needed.

FIGURE 2-3 Categories of dredging and sediment removal equipment. Source: Francingues and Palermo 2006.

Dredging—One Part of the Overall Process Train

Physically removing sediments by dredging is only one component of the overall remediation process. The key processing steps shown in Figure 2-4 include (EPA 2005a):

- Mobilization and setup of equipment.

- Site preparation including debris removal and protection of structures.
- Removal (environmental dredging).
- Staging, transport, and storage (rehandling).
- Treatment (pretreatment, solidification and stabilization of solids, treatment of decant water and/or dewatering effluents and sediment, and potentially separate handling and treatment of materials with and without special requirements under the Toxic Substances Control Act [TSCA]).
- Disposal (liquids and solids).

Environmental dredging must be compatible with all later steps in the process train. For example, the production rate of a dredge (either mechanical or hydraulic) depends heavily on the mode of transportation and the ability to rehandle or directly manage the dredged material on the other end of the process. Compatibility must be considered with respect to the type of pretreatment, treatment, and disposal being planned, especially the availability, size, and capacity of disposal sites, the distance from dredging site to treatment or disposal sites, and constraints associated with production rates for transport, storage, rehandling, treatment, or disposal. Inefficiencies in remedial dredging projects can result from constraints associated with components of the remedy other than dredging, such as dewatering capacity, water-treatment effectiveness, and disposal location and capacity (Palermo et al. 2006).

Dredging accounts for only part of the overall cost of an environmental-dredging project. In a complex project, large costs may be associated with the transport, dewatering, and ultimate disposition of the dredged material. Recent data, described below, support the premise that dredging accounts for 10-20% of the total cost of an environmental dredging project. For example, EPA Region 1 (EPA 2005d) reported on the costs of the 2004 New Bedford Harbor Superfund dredging project,

FIGURE 2-4 Dredging-process train. Source: Adapted from Palermo et al. 2006.

as shown in Figure 2-5; dredging itself represents only 17% of the total yearly construction and operations cost. Similarly, in the Head of Hylebos remediation (Figure 2-6) dredging operations conducted from 2003 to 2006 (Dalton, Olmsted & Fuglevand, Inc. 2006), dredging represents 17% of the total cost. Dredging cost varies widely, depending on many factors, including site conditions, the nature of the sediments and contaminants, the type and size of dredge selected, production rates, and seasonal construction windows. However, when dredging is selected as a remedy, all the other components of the process train will probably be required and will account for most of the overall cost. Only in those cases where transportation and disposal of sediments are relatively inexpensive (for example, where there is an existing in-water or upland disposal site, both suitable for the long-term containment of contaminated sediment and in close proximity to the dredging) will the dredging be a major cost element.

TECHNICAL ISSUES ASSOCIATED WITH DREDGING

Environmental dredging typically strives to achieve contaminant-specific cleanup levels set at each site. A number of technical issues can limit ability or efficiency in achieving those levels. This section describes several of the issues, many of which are revisited in the context of site evaluations in Chapter 4.

Accuracy of Dredging vs. Accuracy of Sediment Characterization

The benefits of being able to position a dredge cut accurately may be achieved only if a corresponding degree of accuracy is reflected in the site and sediment-characterization data. The ability to map the precise location of chemical concentrations accurately both horizontally and vertically depends on the data density (grid density), accessibility of deeper sediments, and other aspects of site characterization (Palermo et al. 2006). In some cases, the ability to locate the dredge cut accurately exceeds the accuracy of the knowledge of the location of the contaminated sediments (Palermo et al. 2006).

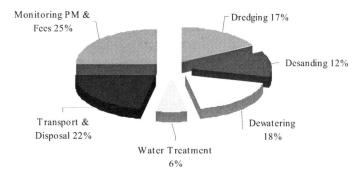

FIGURE 2-5 Cost breakdown of components of environmental dredging at the New Bedford Harbor (New Bedford, MA) project in 2004. Full-scale operations cost about $800,000 per week. PM = project management. Source: Data from EPA 2005d.

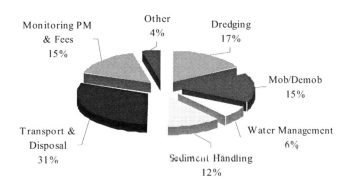

FIGURE 2-6 Cost breakdown of components of environmental dredging at the head of the Hylebos Waterway (Commencement Bay, WA) project. PM = project management; Mob/Demob = mobilization and demobilization. Source: Data from P. Fuglevand, Dalton, Olmsted & Fuglevand, Inc., personal commun., May 11, 2007.

In the context of this discussion, vertical operating accuracy is the ability to position the dredgehead at a desired depth or elevation for the cut, whereas vertical precision is the ability to maintain or repeat the vertical position during dredging. The key to the success of an environmental-dredging project is the removal of the target layer, which is delineated by the cut line, without unnecessary removal of clean material (Palermo et al. 2006).

The ability to dredge to a specified cut line in the sediment has been greatly improved by the advent of electronic positioning technologies, such as differential global positioning systems (DGPSs) and kinematic differential global positioning systems (KDGPSs). Depending on site conditions, dredge operator ability, size and type of dredge, and positioning instrumentation and software, the dredgehead and cut elevation may be locatable with vertical accuracy of less than 30 cm. Vertical accuracies of 10 cm for fixed-arm dredgeheads should be consistently attainable, whereas vertical accuracies of 15 cm should be attainable with proper operator training in the use of wire-supported buckets (Palermo et al. 2006).

Notwithstanding the previous statements regarding accuracies and positioning of dredging equipment, there are numerous challenges in attaining them. For example,

• Effective use of sophisticated dredge positioning systems requires sophisticated operators and contractors in order to achieve the stated accuracies.

• In order to get effective positioning with any of the software packages, the operators must be specifically trained and capable of system operation, and the systems must be properly operated and calibrated.

• Experience has shown that some systems are more difficult to operate than others, and some systems may experience difficulties maintaining calibration. Simply using an electronic positioning system on a dredge does not guarantee that the stated accuracy will be achieved.

Resuspension, Residuals, and Release of Contamination

All dredging equipment disturbs sediment and resuspends some fraction of it in the water column. Resuspended sediment and the associated contaminants can settle back to the bottom in the dredge cut; finer-grained materials can remain in the water column and be transported to other locations. Those materials are deposited as residuals and result from dredging. Dissolved contaminants may also be released to the water column during dredging from resuspended or exposed contaminated sediment. Figure 2-7 is a conceptual illustration of environmental dredging and those processes.

Dredged sediment resuspension, release, and residual and the resulting risk (the "4 Rs") were the focus of a recent workshop held at the U.S. Army Engineer Research and Development Center in Vicksburg, MS (Bridges et al. in press). Effective remediation by dredging requires minimizing the 4 Rs while maximizing the fifth R, removal—either the dredging production rate or the volume removed (Francingues and Thompson 2006). The type and amount of sediment resuspension, contaminant release, and residuals during a dredging operation depend on many site-specific project factors, as shown in Box 2-8.

Resuspension

Resuspension is the process by which dredging and associated operations result in the dislodgement of embedded sediment particles, which disperse into the water column. Resuspended particles may settle in the dredging area or be transported downstream. Recent EPA guidance for sediment remediation states that

> When evaluating resuspension due to dredging, it generally is important to compare the degree of resuspension to the natural sediment resuspension that would continue to occur if the contaminated sediment was not dredged, and the length of time over which increased dredging-related suspension would occur.... Some contaminant release and transport during dredging is inevitable and should be factored into the alternatives evaluation and planned for in the remedy design.... Generally, the project manager should assess all causes of resuspension and realistically predict likely contaminant releases during a dredging operation (EPA 2005a, pp. 6-21, 6-22).

Resuspension concerns related to dredging include the physical effects of turbidity and burial that can result in seasonal restrictions on dredging operations (dredging windows). Sediment resuspension can

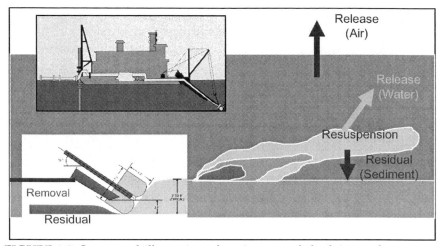

FIGURE 2-7 Conceptual illustration of environmental dredging and processes. Source: Patmont 2006. Reprinted with permission; copyright 2006, Anchor Environmental LLC.

result in chemical releases to the water column (for example, from pore water displaced from the dredged sediment or by desorption from resuspended sediment particles) and residual contamination on the bottom after dredging. Resuspension can be caused not only by dredging equipment but by propwash of tenders (push boats or tugs used to move equipment) and during rehandling and transport operations, such as filling and overflowing of barges and leaky pipelines. Estimates of resuspension from environmental-dredging projects range up to 10% of the mass of sediment dredged (Patmont 2006). Rates of resuspension depend on equipment, material, operator, and other site-specific factors.

Residuals

Residuals are contaminated sediment that remains after dredging. There are two general types of residuals: generated residuals, contaminated sediment that is dislodged or suspended during dredging and later redeposited within or adjacent to the dredging footprint; and undisturbed residuals, contaminated sediments found at the post-dredge sedi-

---

BOX 2-8 Site-Specific Factors Affecting Resuspension, Release, and Residuals Sediment Physical and Chemical Properties

Sediment Physical and Chemical Properties

- Grain size distribution (for example, percentages of silt, clay, and sand).
- Organic carbon content.
- Amount of sulfides.
- Spatial and vertical distributions of contaminants in the sediment (for example, layering).

Site Conditions

- Water velocity and degree of mixing.
- Water salinity, hardness, alkalinity, and temperature.
- Type of substrate (for example, hardpan, bedrock or soft sediment).
- Type and extent of debris in sediment.
- Weather, such as storms that result in wind and waves.
- Wakes from passing vessels.
- Fluctuations in water elevation.
- Depth and slope of area to be dredged.

Equipment

- Type of dredge (for example, cutterhead pipeline, open or closed bucket, and specialty dredgehead).
- Methods of dredging.
- Skill of operators.
- Extent of tender-boat activity.
- Methods of sediment transport and offloading.

Source: Adapted from Palermo et al. 2006.

---

ment surface that have been uncovered but not fully removed as a result of the dredging operation (Bridges et al. in press).

Residuals may result from incomplete characterization, inaccuracies of dredging, mixing of targeted material with underlying materials during dredging, fallback (dislodged sediment not picked up), and resettlement of resuspended sediments (Palermo et al. 2006). Also contributing to residual contamination are such processes as sloughing of sediment into the dredging cut and sloughing induced by bank or slope failures. Site-specific factors, such as debris or limitation of dredging by bedrock or hardpan can influence the amount of residuals. Box 2-9 describes specific processes during dredging that contribute to residual formation.

The residual contaminant mass is typically limited to the upper few inches of sediment, which is populated and actively processed by sediment-dwelling organisms (although in the case of undisturbed residuals the depth can be substantially greater). That upper layer is subject to erosion and other physical and chemical processes that may promote release into the overlying water because of the entrainment of water into the dredged sediment, which causes physical (decreased consolidation) and chemical (redox) changes in the residuals. Residual contamination may also be attributable to sediment that was not dredged, because of the dredger's failure to meet dredge cutlines (either depth or areal targets) or errors or incompleteness in site characterization that failed to identify appropriate depth and areal extent of contaminated sediment.

Patmont (2006) compiled data on residuals from 12 environmental-dredging projects. Final generated residuals ranged from approximately 2 to 9% (average = 5%)[11] of the mass of contaminant dredged during the last production cut. There is little research on the amount of generated residuals transported outside the dredge prism, but their presence has been documented analytically (EcoChem Inc 2005) and visually with sediment-profile imagery (Baron et al. 2005).

Release

Release is the process by which the dredging operation results in the transfer of contaminants from sediment pore water and sediment particles into the water column or air. Contaminants sorbed to resuspended particles may partition to the water column and be transported downstream in dissolved form along with contaminants in the released

---

[11]More recently, Patmont and Palermo (2007) analyzed a similar (though not identical) dataset and found that final generated residuals ranged from approximately 2% to 9% (average = 4%) of the mass of contaminant dredged during the last production cut.

BOX 2-9 Specific Processes Contributing to the Residual
Layer During Dredging

For mechanical dredging, processes that contribute the residual layer are

• The erosion of sediment from around and within the bucket as it is placed on the bottom, closed, and raised through the water column. The erosion in the water column can be controlled with the use of enclosed buckets. However there can be significant resuspenion of contaminated sediment during the closing of enclosed buckets, as the bucket vents expel sediment at high velocity.

• The overflow of turbid water from the sediment haul barge, controlled with restrictions on barge overflow and associated capture and treatment of the turbid water.

For hydraulic dredging, processes that contribute the residual layer are

• The spillage layer generated by hydraulic dredging associated with the turning of the cutterhead or auger in the sediment. Hydraulic dredges are normally configured with the inlet of the suction pipe well above the lowest reach of the rotating cutterhead or auger. That means that the mixed layer generated by the cutterhead or auger is not fully removed by the suction pipe and consequently there is a "spillage layer" left behind after dredging.

• Another source of residual sediment is resuspension by the rotating cutterhead or auger, when sediment is displaced away from the cutterhead or auger into the water column.

Dredging, either mechanical or hydraulic, can result in the formation of a residual layer through a variety of mechanisms including

• The sloughing of the sidewalls and headwall of the dredge cut face back on to previously dredged areas. This sloughing can be controlled through the use of relatively thin dredge lifts (few feet each) and by including a final cleanup pass of dredging once the bulk of sediment has first been removed ("two pass dredge approach"). If not controlled, this bank sloughing can result in a considerable residual layer forming on previously dredged areas.

• The remolding of soft fine-grained sediment by the dredging process can significantly reduce the strength of the material and generate a more liquid like flowable residual layer in the dredging area. This flowable material can be very difficult to capture with the dredge and result in a residual layer that is

(Continued on next page)

BOX 2-9 Continued

difficult to manage and control once it is formed. The formation of this layer can be reduced (not eliminated) by a controlled and precise removal program using electronic, GPS-enabled dredge positioning and mechanical dredging. Once formed, capture of the flowable layer can be accomplished with overdredging into native substrate, provided that substrate is not hardpan or bedrock.

Sources: Adapted from Dalton, Olmsted & Fuglevand, Inc. 2006; Fuglevand and Webb 2006, 2007; Hartman 2006.

---

pore water. Contaminants in the generated or undisturbed residuals may be released to the water column by densification, diffusion and bioturbation (Bridges et al. in press).

Releases of contaminants from the aforementioned sources and processes are considered to be up to about 5% of the contaminant mass in the sediment dredged, but larger or smaller releases may be observed, depending on site-specific factors and the type and operating characteristics of the dredge (Sanchez 2001; Sanchez et al 2002). The degree of contaminant release to the air and water is directly related to the degree of sediment resuspension (and pore water release) and chemical properties affecting the mass transfer of contaminants. Therefore, control of resuspension should have high priority at many dredging project sites that involve contaminated sediment. Contamination can also be released from sediment beds to the water column in soluble form without particle resuspension (Thibodeaux and Bierman 2003; Erickson et al. 2005). That suggests that the residual layer is also a contributor of contaminant release after dredging. Control of solids is important but is not always sufficient to prevent contaminant losses.

Impact on Risk

Risk can result from contaminant exposures driven by resuspension, production of residuals, and contaminant release. Those processes are important because they can alter the accessibility bioavailability of contaminants, create additional contaminant exposure pathways that potentially affect the risk resulting from dredging, and may continue to influence risk after remedial operations cease. Surface-water concentrations and surface-sediment concentrations may increase during and after dredging and can result in adverse effects and accumulation of contaminants in organisms. The potential for volatile compounds to be released into the air may be an additional concern in connection with highly contaminated sites (EPA 2005a).

Release, resuspension, and production of residuals will affect risk over different spatial scales and time frames depending on the site characteristics and nature of the dredging operation. As described by Bridges et al. (in press), "Characterizing how dredging will influence direct risks includes considering how the processes contributing to risk change with time, which elements or receptors in the ecosystem are affected by these changes, the spatial scales over which effects would be expected to occur, and the uncertainties associated with the predicted changes and risk reduction." As will be discussed in much greater detail in Chapter 4, resuspension and release occur in a shorter time frame during dredging operations. Residuals will remain following dredging, however, their distribution, longevity, and effects are poorly understood. To the extent that release, resuspension, and production of residuals are present and contribute risk at a site, they detract from the overall or net risk reduction resulting from the remedial activity. As such, they are an important consideration in evaluating the effectiveness of a remediation. As noted in the 4 Rs workshop (Bridges et al. in press) and recent EPA sediment guidance (EPA 2005a), there is increasing recognition of the importance of these processes and of factors that influence their control.

REFERENCES

Averett, D.E., B.D. Perry, E.J. Torre, and J.A. Miller. 1990. Review of Removal, Containment, and Treatment Technologies for Remediation of Contaminated Sediments in the Great Lakes. Miscellaneous Paper EL-90-25. Prepared for Great Lakes National Program Office, U.S. Environmental Protection Agency, Chicago, IL, by U.S. Army Corps of Engineers Waterways Experiment Station, Vicksburg, MS.

Baron, L.A., M.R. Bilimoria, and S.E. Thompson. 2005. Environmental dredging pilot study successfully completed on the Lower Passaic River, NJ—one of America's most polluted rivers. World Dredging Mining and Construction

(Dec. 2005):13-18 [online]. Available: http://www.ourpassaic.org/project sites/premis_public/home/caPassaicPullout_final.pdf [accessed Jan. 3, 2007].

Bohlen, W.F., and M.J. Erickson. 2006. Incorporating sediment stability within the management of contaminated sediment sites: A synthesis approach. Integr. Environ. Assess. Manage. 2(1):24-28.

Bray, R.N. 1979. Dredging: A Handbook for Engineers. London: Arnold.

Bridges, T.S., S. Ells, D. Hayes, D. Mount, S. Nadeau, M. Palermo, C. Patmont, and P. Schroeder. in press. The Four Rs of Environmental Dredging: Resuspension, Release, Residual, and Risk. ERDC TR-07-X. U.S. Army Engineer Research and Development Center, Vicksburg, MS.

Dalton, Olmsted & Fuglevand, Inc. 2006. Remediation Action Construction Report, Part 1: Head of Hylebos Waterway Problem Area Commencement Bay Nearshore/Tideflats Superfund Site Tacoma, Washington, Review Draft, July 21, 2006. Prepared for Head of Hylebos Cleanup Group, Arkema, Inc, General Metals of Tacoma, Inc, by Dalton, Olmsted & Fuglevand, Inc., Kirkland, WA. July 21, 2006.

EcoChem, Inc. 2005. Duwamish/Diagonal CSO/SD Sediment Remediation Project Closure Report. Prepared for King County Department of Natural Resources and Parks, Elliott Bay/Duwamish Restoration Program Panel, Seattle, WA. Panel Publication 39. July 2005 [online]. Available: http://dnr. metrokc.gov/WTD/duwamish/diagonal.htm [accessed Jan. 3, 2006].

EPA (U.S. Environmental Protection Agency). 1988. Guidance for Conducting Remedial Investigations and Feasibility Studies under CERCLA, Interim Final. EPA 540/G-89/004. OSWER 9355.3-01. Office of Emergency and Remedial Response, U.S. Environmental Protection Agency, Washing-ton, DC [online]. Available: http://www.epa.gov/superfund/resources/remedy/ pdf/540g-89004.pdf [accessed Jan. 11, 2005].

EPA (U.S. Environmental Protection Agency). 1994. Assessment and Remediation of Contaminated Sediments (ARCS) Program. Pilot-Scale Demonstration of Solvent Extraction for the Treatment of Grand Calumet River Sediments. EPA/905/R-94/003. Great Lakes National Program Office, U.S. Environmental Protection Agency, Chicago, IL. January 1994.

EPA (U.S. Environmental Protection Agency). 1997. The Incidence and Severity of Sediment Contamination in Surface Waters of the United States: Report to Congress, Vols. 1-3. EPA 823-R-97-006, EPA 823-R-97-007, EPA 823-R-97-008. Office of Science and Technology, U.S. Environmental Protection Agency, Washington, DC. September 1997 [online]. Available: http://www. epa.gov/waterscience/cs/congress.html [accessed Dec. 26, 2006].

EPA (U.S. Environmental Protection Agency). 1998. EPA's Contaminated Sediment Management Strategy. EPA-823-R-98-001. Office of Water, U.S. Environmental Protection Agency. April 1998 [online]. Available: http://www. epa.gov/waterscience/cs/strategy.pdf [accessed Dec. 26, 2006].

EPA (U.S. Environmental Protection Agency). 2000. Close Out Procedures for National Priorities List Sites. EPA 540-R-98-016. Office of Emergency and Remedial Response, U.S. Environmental Protection Agency, Wash-ington, DC. January 2000 [online]. Available: http://www.epa.gov/superfund/ resources/closeout/pdf/guidance.pdf [accessed Aug. 3, 2006].

EPA (U.S. Environmental Protection Agency). 2001. Comprehensive Five-Year Review Guidance. EPA 540-R-01-007. Office of Emergency and Remedial Response, U.S. Environmental Protection Agency, Washing-ton, DC. June 2001 [online]. Available: http://www.epa.gov/superfund/resources/5year/ guidance.pdf [accessed Aug. 3, 2006].

EPA (U.S. Environmental Protection Agency). 2003a. Urban Rivers Restoration Initiative. EPA 500-F-03-023. Office of Solid Waste and Emergency Response, U.S. Environmental Protection Agency, Washington, DC. October 2003. [online]. Available: http://www.epa.gov/oswer/land revitalization/download/urri_brochure.pdf [accessed Aug. 3, 2006].

EPA (U.S. Environmental Protection Agency). 2003b. Procedures for the Derivation of Equilibrium Partitioning Sediment Benchmarks (ESBs) for the Protection of Benthic Organisms: PAH Mixtures. EPA/600/R-02/013. Office of Research and Development, U.S. Environmental Protection Agency, Washington DC [online]. Available: http://www.epa.gov/nheerl/ publications/files/PAHESB.pdf [accessed May 1, 2007].

EPA (U.S. Environmental Protection Agency). 2004a. Congressional Request on Funding Needs for Non-Federal Superfund Sites. Report 2004-P-00001. Office of Inspector General, U.S. Environmental Protection Agency, Washington, DC. January 7, 2004. [online]. Available: http://www.epa. gov/oig/reports/2004/20040107-2004-p-00001.pdf [accessed Jan. 18, 2007].

EPA (U.S. Environmental Protection Agency). 2004b. The Incidence and Severity of Sediment Contamination in Surface Waters of the United States: National Sediment Quality Survey, 2nd Ed. EPA-823-R-04-007. Office of Science and Technology, U.S. Environmental Protection Agency, Washing-ton, DC. November 2004 [online]. Available: http://www.epa.gov/water science/cs/report/2004/nsqs2ed-complete.pdf [accessed Dec. 26, 2006].

EPA (U.S. Environmental Protection Agency). 2004c. Great Lakes Legacy Act of 2002. Fact Sheet. Great Lakes National Program Office, U.S. Environmental Protection Agency. January 2004 [online]. Available: http://www.epa.gov/ glnpo/sediment/legacy/glla-factsheet-200401.pdf [accessed Aug. 3, 2006].

EPA (U.S. Environmental Protection Agency). 2005a. Contaminated Sediment Remediation Guidance for Hazardous Waste Sites. EPA-540-R-05-012. OSWER 9355.0-85. Office of Solid Waste and Emergency Response, U.S. Environmental Protection Agency. December 2005 [online]. Available:

http://www.epa.gov/superfund/resources/sediment/pdfs/guidance.pdf [accessed Dec. 26, 2006].

EPA (U.S. Environmental Protection Agency). 2005b. 2004 National Listing of Fish Advisories. Fact Sheet. EPA-823-F-05-004. Office of Water, U.S. Environmental Protection Agency. September 2005 [online]. Available: http://www.epa.gov/waterscience/fish/advisories/fs2004.pdf [accessed Jan. 3, 2006].

EPA (U.S. Environmental Protection Agency). 2005c. Procedures for the Derivation of Equilibrium Partitioning Sediment Benchmarks (ESBs) for the Protection of Benthic Organisms: Metal Mixtures (Cadmium, Copper, Lead, Nickel, Silver and Zinc). EPA/600/R-02/011. Office of Research and Development, Washington DC [online]. Available: http://www.epa.gov/nheerl/publications/files/metalsESB_022405.pdf [accessed May 1, 2007].

EPA (U.S. Environmental Protection Agency). 2005d. Sediment Remediation Course: Technical Considerations for Evaluating and Implementing Dredging and Capping Remedies, August 16-18, 2005, Boston, MA

EPA (U.S. Environmental Protection Agency). 2006a. Contaminated Sediment in Superfund. U.S. Environmental Protection Agency [online]. Available: http://www.epa.gov/superfund/resources/sediment/ [accessed Jan. 3, 2006].

EPA (U.S. Environmental Protection Agency). 2006b. List of Sediment Sites with Substantial Contamination. Contaminated Sediment in Superfund. U.S. Environmental Protection Agency [online]. Available: http://www.epa.gov/superfund/resources/sediment/sites.htm [accessed Jan. 3, 2006].

Erickson, M.J., C.L. Turner, and L.J. Thibodeaux. 2005. Field observation and modeling of dissolved fraction sediment-water exchange coefficients for PCBs in the Hudson River. Environ. Sci. Technol. 39(2):549-556.

Fletcher, S.R., M. Isler, D.M. Bearden, C. Copeland, R. Esworthy, L. Luther, J.E. McCarthy, J.L. Ramseur, M. Reisch, L.J. Schierow, M. Tiemann, and B.D. Yacobucci. 2006. Environmental Protection Issues in the 109th Congress. CRC Report for Congress. CRS Web Order Code RL33481. Congressional Research Service. June 23, 2006 [online]. Available: http://fpc.state.gov/documents/organization/68817.pdf [accessed Dec. 8, 2006].

Francingues, N.R., and D.W. Thompson. 2006. Control of resuspended sediments in dredging projects. Pp. 243-254 in Proceedings of the Western Dredging Association 26th Technical Conference and 38th Annual Texas A&M Dredging Seminar, June 25-26, 2006, San Diego, CA, R.E. Randell, ed. Center for Dredging Studies, Ocean Engineering Program, Civil Engineering Department, Texas A&M University, College Station, TX.

Francingues, N.R., and M.R. Palermo. 2006. Dredging Overview. Presentation at the 16th NARPM-National Association of Remedial Project Managers Annual Training Conference, June 21, New Orleans, LA.

Fuglevand, P.F., and R.S. Webb. 2006. Water management during mechanical dredging of contaminated sediment. Pp. 461-467 in Proceedings of the Western Dredging Association 26th Technical Conference and 38th Annual Texas A&M Dredging Seminar, June 25-26, 2006, San Diego, CA, R.E. Randell, ed. Center for Dredging Studies, Ocean Engineering Program, Civil Engineering Department, Texas A&M University, College Station, TX.

Fuglevand, P.F., and R.S. Webb. 2007. Head of Hylebos—Adaptive Management during Sediment Remediation. Preprint for the World Dredging Congress (WODCON XVIII): Global Dredging: Its Impact on the Economy and Environment, May 27-June 1, 2007, Orlando, FL.

Hartman, G.L. 2006. Hydraulic Pipeline Dredge Evaluation For Remedial Dredging. Expert Panel on Sediment Remediation, November 1-2, 2006, Chicago, IL.

Herbich, J.B. 2000. Handbook of Dredging Engineering, 2nd Ed. New York: McGraw-Hill.

Menozzi, R.L., E. Ould, G. Green, C. Moses, T. Binsfeld, S. McGee, V. Buhr, and T. Bumstead. 2003. Grand Calumet River sediment remediation project: Largest impacted sediment, hydraulically dredged project in North America. M-01 in Proceedings of the Second International Conference on Remediation of Contaminated Sediments, M. Pellei, and A. Porta, eds. Columbus, OH: Battelle Press.

NRC (National Research Council). 1997. Contaminated Sediments in Ports and Waterways. Washington, DC: National Academy Press.

NRC (National Research Council). 2001. A Risk-Management Strategy for PCB-Contaminated Sediments. Washington, DC: National Academy Press.

NRC (National Research Council). 2003. Bioavailability of Contaminants in Soils and Sediments. Washington, DC: National Academies Press.

Palermo, M.R., S. Maynord, J. Miller, and D.D. Reible. 1998. Guidance for In-Situ Subaqueous Capping of Contaminated Sediments. EPA 905-B96-004. Great Lakes National Program Office, Chicago, IL [online]. Available: http://www.epa.gov/glnpo/sediment/iscmain/index.html [accessed Jan. 3, 2006].

Palermo, M.R., N.R. Francingues, and D. E.Averett. 2003. Equipment selection factors for environmental dredging. Pp 342-363 in Proceedings of the Western Dredging Association 23rd Technical Conference and 35th Annual Texas A&M Dredging Seminar, June 10-13, Chicago, IL, R.E. Randell, ed. CDS Report No. 376. Center for Dredging Studies, Ocean Engineering Program, Civil Engineering Department, Texas A&M Uni-versity, College Station, TX.

Palermo, M.R., N.R. Francingues, and D.E. Averett. 2004. Operational Characteristics and Equipment Selection Factors for Environmental Dredging. Journal of Dredging Engineering, Western, Dredging Association, Vol. 5, No. 4.

Palermo, M. R., N.R. Francingues, P.R. Schroeder, and T.O. Estes. 2006. Guidance
    for Environmental Dredging of Contaminated Sediments, Draft. DOER TR-
    X-X. Prepared for Office of Solid Waste and Emergency Response, Wash-
    ington, DC, by U.S. Army Engineer Research and Development Center,
    Vicksburg, MS.
Patmont, C. 2006. Contaminated Sediment Dredging Residuals: Recent Monitor-
    ing Data and Management Implications. Presentation at the Second Meet-
    ing on Sediment Dredging at Superfund Megasites, June 7, 2006, Irvine,
    CA.
Patmont, C., and M. Palermo. 2007. Case Studies Environmental Dredging Re-
    siduals and Management Implications. Paper D-066 in Remediation of
    Contaminated Sediments-2007: Proceedings of the 4th International Con-
    ference on Remediation of Contaminated Sediments, January 2007, Savan-
    nah, GA. Columbia, OH: Battelle Press.
Patmont, C., J. Davis, T. Dekker, M. Erickson, V.S. Magar, and M. Swindoll. 2004.
    Natural recovery: Monitoring declines in sediment chemical concentrations
    and biological endpoints. Paper No. C-04 in Remediation of Contaminated
    Sediments-2003: Proceedings of the Second International Conference on
    Remediation of Contaminated Sediments, Sept. 30-Oct. 3, 2003, Venice, It-
    aly, M. Pellei, and A. Porta, eds. Columbus, OH: Battelle Press.
PCCRARM (Presidential/Congressional Commission on Risk Assessment and
    Risk Management). 1997. Risk Assessment and Risk Management in
    Regulatory Decision-Making, Vol. 2. Washington, DC: U.S. Government
    Printing Office [online]. Available: http://www.riskworld.com/Nreports/
    1997/risk-rpt/volume2pdf/v2epa.pdf [accessed Oct. 3, 2006].
Probst, K.N., D. Fullerton, R.E. Litan, and P.R. Portney. 1995. Footing the Bill for
    Superfund Cleanups Who Pays and How? Washington, DC: Resources for
    the Future.
Probst, K.N., D.M. Konisky, R. Hersh, M.B. Batz, and K.D. Walker. 2001.
    Superfund's Future: What Will It Cost? A Report to Congress. Washing-
    ton, DC: Resources for the Future [online]. Available: http://www.rff.org/
    rff/RFF_Press/CustomBookPages/Superfunds-Future.cfm?CFID=5819288&
    CFTOKEN=75932196 [accessed Jan. 9, 2006].
Sanchez, F.F. 2001. A Multimedia Model for Assessing Chemical Fate During
    Dredging of Contaminated Bed-Sediment. M.S. Thesis, Louisiana State
    University.
Sanchez, F.F., L.J. Thibodeaux, K.T. Valsaraj, and D.D. Reible. 2002. Multimedia
    Chemical Fate Model for Environmental Dredging. Pract. Period. Hazard.
    Tox. Radioact. Waste Manage. 6(2):120-128.
Southerland, E. 2006. Overview of Superfund Sediment Remediation. Presenta-
    tion at the First Meeting on Sediment Dredging at Superfund Megasites,
    March 22, 2006, Washington DC.

Thibodeaux, L.J., and V.J. Bierman. 2003. The bioturbation-driven chemical release process. Environ. Sci. Technol. 37(13):252A-258A.

Thoms, S.R., G. Matisoff, P.L. McCall, and X. Wang. 1995. Models for alteration of sediments by benthic organisms. Project 92-NPS-2. Alexandria, VA: Water Environment Research Foundation.

USACE (U.S. Army Corps of Engineers). 1983. Dredging and Dredged Material Disposal. Engineer Manual 1110-2-5025. U. S. Army Corps of Engineers, Department of the Army, Washington, DC. March 25, 1983 [online]. Available: http://www.usace.army.mil/publications/eng-manuals/em11102 5025/entire.pdf [accessed Jan. 4, 2006].

USACE /EPA (U.S. Army Corps of Engineers and U.S. Environmental Protection Agency). 1992. Evaluating Environmental Effects of Dredged Material Management Alternatives - A Technical Framework. EPA842-B-92-008. U.S. Army Corps of Engineers, Department of the Army, and Office of Water, U.S. Environmental Protection Agency, Washington, DC [online]. Available: http://www.epa.gov/owow/oceans/regulatory/dumpdredged/ framework/techframework.pdf [accessed Jan. 4, 2006].

U.S. Steel. 2004. Grand Calumet River Sediment Remediation Project Update. U.S. Steel Gary Works Newsletter GCR Issue 6 - February 2004 [online]. Available: http://www.ussteel.com/corp/rcra/images/CITE/0402%20News letter%20Feb%2004.pdf [accessed Aug. 3, 2006].

Washington Department of Ecology. 2005. Sediment Cleanup Status Report. Washington State Department of Ecology Toxics Cleanup Program. Publication No. 05-09-092. June 2005 [online]. Available: http://www.ecy.wa. gov/pubs/0509092.pdf [accessed Feb. 26, 2007].

Wenning, R.J., M. Sorensen, and V.S. Magar. 2006. Importance of implementation and residual risk analyses in sediment remediation. Integr. Environ. Assess. Manage. 2(1):59-65.

Weston (Weston Solutions Inc.). 2002. Final Comprehensive Post-Removal Summary Report Manistique Harbor and River Site U.S. Route 2 Highway Bridge Manistique, Michigan. Prepared for the U.S. Environmental Protection Agency, Region V. November 12, 2002.

3

Effectiveness of Environmental Dredging in Reducing Risk: Framework for Evaluation

A wide variety of metrics can be used to evaluate the effectiveness of environmental-dredging projects in reducing risks to human health and the environment. The committee reviewed a number of them and developed a framework to facilitate the evaluation of the effectiveness of environmental-dredging projects at contaminated sediment sites. The framework is based on the effectiveness criteria used by the U.S. Environmental Protection Agency (EPA) Superfund program to select remedies (40 CFR § 300.430[e][7][i]).

In conducting its evaluation, the committee defined dredging effectiveness as the achievement of cleanup goals defined for each site, which take the form of remedial-action objectives, remediation goals, and cleanup levels.[1] For CERCLA remedial actions, these goals are typically

---

[1]*Remedial-action objectives* "are intended to provide a general description of what remediation is expected to accomplish" (EPA 2005a, p. 2-15). For example, a remedial-action objective might be to reduce to acceptable levels the risks to people who ingest contaminated fish. *Remediation goals* are paired contaminant-specific and media-specific concentrations intended to protect human and ecologic health that incorporate site-specific information about exposure patterns and toxicity. "At most CERCLA [Comprehensive Environmental Response, Compensation, and Liability Act] sites, [remediation goals] for human health and ecologic

documented in the record of decision (ROD). However, this definition is appropriate for evaluating the effectiveness of the project in reducing risk only when cleanup goals are derived from sound, site-specific risk modeling.

An ideal evaluation of effectiveness at Superfund sites would be based on the site conceptual model, data from a baseline assessment, and a long-term monitoring program that permits sound statistical comparison of the spatial scale of contamination and the magnitude of risk before, during, and after dredging. At the outset of its work, the committee hoped to obtain that kind of information for a number of large contaminated sediment sites to inform its deliberations. However, we found that such careful and prolonged monitoring either has not been conducted, has not been completed at large-contaminated sediment sites, or simply was not available to the committee. (The committee noted that some sites where remediation had not yet occurred or been completed have electronic databases and long-term monitoring plans that would facilitate future attempts to evaluate remedy effectiveness, for example, Hudson River and New Bedford Harbor). In some cases, it is recognized that additional information, for example, raw data and consulting reports, may have been held by responsible parties, federal agencies, or their consultants. However, this information may not have been available in the public domain or to the committee, or the committee's time and resource constraints precluded a thorough compilation, analysis, and interpretation (see further discussion in Chapter 4). In some cases a review of all site data was not necessary to determine whether cleanup goals had been met. This chapter details the committee's process for evaluating the effectiveness of environmental dredging within the constraints imposed by the available data. The framework for this review is outlined in Box 3-1.

---

receptors are developed into final, chemical-specific, sediment cleanup levels by weighing a number of factors, including site-specific uncertainty factors and the criteria for remedy selection found in the NCP [National Contingency Plan]" (EPA 2005a, p. 2-16). The ROD for each site generally should include chemical-specific cleanup levels, indicating how these values are related to risk and how their attainment will be measured (EPA 2005a).

BOX 3-1 Framework for Evaluating the Effectiveness of
Environmental Dredging Projects

    1.   Identify Superfund megasites and other large environmental-dredging projects.

    2.   Define criteria for selecting projects for committee evaluation from the list of environmental-dredging projects.

    3.   Select projects that represent a variety of site conditions.

    4.   Evaluate each project with respect to measures of short-term and long-term effectiveness.

    5.   Make recommendations for improved design, implementation, and monitoring of future environmental-dredging projects.

SOURCES OF INFORMATION ABOUT ENVIRONMENTAL-DREDGING PROJECTS

The statement of task (see Appendix A) indicates that the sources of information "would include megasites for which dredging has been completed; megasites for which plans have been developed, partially implemented, and operations are ongoing; and smaller sites that exhibit lessons relevant to megasites." The committee's evaluation focused on environmental-dredging projects at Superfund megasites, but, because remediation has not been completed at many of these sites, the committee also reviewed other environmental-dredging projects on which data relevant to the committee's charge were available.

As described in Chapter 2, numerous environmental-dredging projects have been conducted, and many other contaminated sediment sites will require decision-making soon. At the first committee meeting, EPA outlined sites from its database of tier 1 sediment sites where dredging had been completed. From those dredging sites, EPA provided the committee with a list of sites on which there were pre-remediation and post-remediation monitoring data. To identify other dredging projects for possible evaluation, the committee reviewed information from additional government, industry, and private consulting sources that summarize remedial activities at contaminated sediment sites (see Box 3-2). Collectively, those dredging project compilations and reviews

BOX 3-2 Compilations and Reviews of Sediment Remediation Projects

EPA's Great Lakes National Program Office summarized (EPA 1998) and updated (EPA 2000) information on contaminated sediment sites in the Great Lakes Basin that had been remediated with a variety of techniques. The reports provide some details about remedies but few details on post-remedial monitoring or on whether remediation achieved expected benefits. In 1999, the Great Lakes Water Quality Board summarized 20 sediment remediation projects in the Great Lakes areas of concern (Zarull et al. 1999). The report stated that of the projects implemented so far, only two (Waukegan Harbor, IL and Black River, OH) had adequate data on long-term ecologic health. The Great Lakes binational strategy progress reports (e.g., EPA and Environment Canada 2005) provide annual updates summarizing sediment remediation activities at Great Lakes sites in the United States and Canada. The Major Contaminated Sediment Sites (MCSS) Database (GE et al. 2004) is the largest compendium of information on such sites. It contains information on 123 major projects representing 103 sites. EPA maintains, although not publicly, a database on the 60 tier 1 sediment sites where remedies include dredging or excavation of at least 10,000 cy or capping or monitored natural recovery of at least 5 acres (EPA 2005a).

Before this committee's deliberations, several other groups reviewed data on completed projects to examine whether remediation had been successful and to draw conclusions about the likely effects of dredging at other sites in the future. The General Electric Company (GE 2000) evaluated sediment remediation case studies involving 25 sites (including sites that used removal techniques other than dredging) and attempted to determine whether data indicated the ability to reduce risks to human health and the environment. Its report concluded that the success of the projects had not been demonstrated and that technical limitations restricted the effectiveness of dredging in reducing surface sediment-contaminant concentrations. Cleland (2000) updated an earlier report prepared by Scenic Hudson (2000) and outlined experience at 15 sediment remediation sites, including many evaluated by GE (2000). Cleland presented sediment and biota concentration data and concluded that post-dredging monitoring data consistently show beneficial results, including reductions in contaminant mass and in concentrations in sediment and fish. Many of the same case studies were also reviewed by EPA (Hahnenberg 1999), and EPA presented the results of its analysis to the National Research Council's Committee on Remediation of PCB-Contaminated Sediments (NRC 2001). EPA's analysis indicated that dredging resulted in reduced sediment and fish-tissue concentrations. An alternative analysis was also provided to that National Research Council committee by the

(Continued on next page)

BOX 3-2 Continued

Fox River Group (1999, Appendix C in GE 2000), which refuted the connection between remedial actions and fish-tissue concentration declines on the basis of a paucity of data and flawed method in EPA's linking of remediation to observed changes. During its deliberations, the previous National Research Council committee (NRC 2001) reviewed the documents produced by GE (2000), Scenic Hudson (2000), EPA (Hahnenberg 1999; Pastor 1999), and the Fox River Group (1999) to ascertain how groups reviewing the same documents could come to such disparate conclusions. It concluded: "First, in some instances, there is disagreement about the remediation goals and the measures by which achievement of the goals can be assessed. Second, in some cases, the available post-remediation monitoring data are sparse and incomplete compared with pre-remediation data and control data. Third, in some cases, it is the intention of reviewers, agencies, and industries to support their preferences, and that might lead to more conflict."

Since the previous committee issued its report (2001), additional groups have sought to analyze the link between sediment remediation and reduced risk. Baker et al. (2001) sought to address the question "Can active remediation be implemented in such a way that it provides a net benefit to the Hudson River?" and commented on the biologic effects of sediment remediation at five sites. Thibodeaux and Duckworth (2001) evaluated measurements of environmental-dredging effectiveness in detail at three sites. Malcolm Pirnie (Malcolm Pirnie and TAMS Consultants, Inc. 2004), in an appendix to the engineering performance standards developed for the Hudson River Superfund site, briefly reviewed data on 25 remediation projects (some had not initiated remediation) and provided information on monitoring and remediation results. As part of its feasibility study for the Kalamazoo River in Michigan, BBL (2000) compiled profiles of 20 environmental-dredging projects conducted nationwide and discussed their ability to meet project-specific objectives and their overall effectiveness. Several of those profiles were presented in an earlier analysis (GE 2000). In addition, the Great Lakes Dredging Team, a partnership of federal and state agencies, provided a series of dredging project case studies (Great Lakes Dredging Team 2006). The U.S. Army Corps of Engineers Center for Contaminated Sediments also provided information on a series of environmental-dredging projects (USACE 2006).

---

were extremely useful in forming the short list of dredging projects on which pre-dredging and post-dredging monitoring data were likely to be suitable for evaluation.

CRITERIA USED TO SELECT ENVIRONMENTAL-DREDGING
PROJECTS FOR EFFECTIVENESS EVALUATION

The committee used the criteria listed in Box 3-3 to select environmental-dredging projects for evaluation. Therefore, the committee preferred projects that involved only dredging but did not limit its analysis to them. The committee did not select projects that involved dry excavation, because they are conducted under different conditions from conventional dredging. Pilot studies, although limited in scope and purpose, were also considered because they are often heavily monitored with substantial information regarding the effect of dredging under specified conditions.

The committee preferred projects with removal of at least 10,000 cy of sediment because of their similarity to megasites with respect to the spatial scale of ecologic and human health exposures. Some smaller pilot studies designed to inform decisions on larger sites were also included. This size threshold matches the threshold used by EPA to define tier 1 sediment sites (EPA 2005a).

Any evaluation of dredging effectiveness depends on sufficient high-quality pre-dredging and post-dredging monitoring data. Therefore, another criterion used by the committee to select dredging projects was whether they had pre-dredging and post-dredging monitoring data.

Environmental-dredging projects occur in a wide variety of aquatic environments, such as rivers, estuaries, bays, lakes, ponds, canals, and wetlands; and the effectiveness of dredging can be influenced by many

---

BOX 3-3 Criteria Used to Select Environmental-Dredging
Projects for Evaluation

1.   The remedy consists of dredging only or a combined remedy that includes dredging.
2.   The remedy preferably includes removal of at least 10,000 yds^3 of sediment.
3.   The project has some amount of pre-dredging and post-dredging data.
4.   The projects collectively represent a wide variety of project types (for example, environmental settings, chemicals of concern, and dredging design and implementation).

site-specific factors. Therefore, the committee endeavored to conduct a broad review of projects that represented a variety of site-specific conditions, including the type of water body, the form of chemical contamination, and the type of dredging technology.

SELECTION OF ENVIRONMENTAL-DREDGING PROJECTS

The committee chose the 26 dredging projects listed in Table 3-1 for detailed evaluation (see Figure 3-1 for the location of the projects). Selected projects involved remedies with dredging only or dredging combined with backfilling or capping. Chemicals of concern included polychlorinated biphenyls (PCBs), polycyclic aromatic hydrocarbons (PAHs), pesticides, and metals detected in sediments in a variety of aquatic environments across the United States and one in Sweden. PCBs were the primary chemicals of concern in 19 of the 26 projects. Table 3-1 contains general information on the site and the remedy; further detail on several of the sites is provided in Chapter 4 and the associated references. In addition, summaries of many of these sites and their remediation have been compiled elsewhere (e.g., GE 2004; Malcolm Pirnie, Inc. and TAMS Consultants, Inc. 2004).

The selected projects include all 12 sites (although not all the operable units at each site) on which there were pre-dredging and post-dredging data as reported by EPA at the first committee meeting (Ells 2006). The committee also chose three demonstration projects (two in the Lower Fox River, WI, and one in the Grasse River, NY) and one state-lead site (Cumberland Bay, NY) recommended for review by the Sediment Management Work Group during the committee's second meeting (Nadeau 2006).

Scenic Hudson (Cleland 2000), an advocacy group that promotes cleanup of the Hudson River in New York, also included those demonstration projects and Cumberland Bay in its review of dredging effectiveness. In addition, Scenic Hudson identified Lake Jarnsjon in Sweden and the Black River, OH, as examples of successful dredging operations (as did Green and Savitz [unpublished] and Zarull et al. 1999), and the committee selected these two sites for evaluation. During the committee's second meeting, remediation consultants presented results of two recently completed large-scale dredging operations: in the Lower Fox

TABLE 3-1 Summary of 26 Environmental-Dredging Projects Selected for Evaluation[a]

Project	Primary Chemicals of Concern	Water-Body Type	Remedy Type	Scale of Effort	Dates of Dredging	Volume of Dredged Sediment (cy)
Bayou Bonfouca, LA	PAHs	Bayou	Dredging followed by backfilling	Full-scale	1994-1995	170,000
Lavaca Bay, TX	Hg	Estuary	Dredging	Pilot	1998	80,000
Black River, OH	PAHs	River	Dredging	Full-scale	1989-1990	45,000-60,000
Outboard Marine Corp., Waukegan Harbor, IL	PCBs	Harbor	Dredging	Full-scale	1991-1992	38,000
Commencement Bay–Head of Hylebos, Tacoma, WA	PCBs, As, PAHs	Estuary	Dredging	Full-scale	2003-2006[b]	419,000[c]
Commencement Bay–Sitcum, Tacoma, WA	As, Cu	Estuary	Dredging	Full-scale	1993-1994	428,000[d]
Duwamish Diagonal, Seattle, WA	PCBs	Tidally influenced river	Dredging followed by capping	Full-scale	2003-2004	66,000

(Continued on next page)

TABLE 3-1 Continued

Project	Primary Chemicals of Concern	Water-Body Type	Remedy Type	Scale of Effort	Dates of Dredging	Volume of Dredged Sediment (cy)
Puget Sound Naval Shipyard, Bremerton, WA	PCBs	Estuary	Dredging	Full-scale	2000-2004	225,000[c]
Harbor Island–Lockheed, Seattle, WA	PCBs, PAHs, Hg, Pb, As, Cu, Zn, tributyltin	Estuary	Dredging in open-water area, dredging and capping in nearshore area	Full-scale	2003-2004[f]	70,000
Harbor Island–Todd, Seattle, WA	As, Pb, Zn, Cu, PAHs, PCBs, tributyltin, Hg	Estuary	Dredging, capping, and habitat restoration in selected areas	Full-scale	2004-2005	220,000
Cumberland Bay, NY	PCBs	Inland lake	Dredging	Full-scale	1999-2000	195,000
Dupont, Christina River, DE	Zn, Pb, Cd	River	Dredging and backfilling	Full-scale	1999	11,000
Lower Fox River (OU-1), WI	PCBs	River	Dredging	Full-scale	2004-present	Incomplete
Lower Fox River (Deposit N), WI	PCBs	River	Dredging	Pilot	1998-1999	8,200

Site	Contaminant	Water body	Remediation	Scale	Year	Volume
Lower Fox River (SMU 56/57), WI	PCBs	River	Dredging and backfilling	Pilot	1999-2000	82,000
Ketchikan Pulp Company, Ward Cove, AK	4-methyl phenol; ammonia	Fjord	Dredging and backfilling (thin-layer)	Full-scale	2000-2001	8,700
Newport Naval Complex–McCallister Landfill, RI	PAHs, PCBs	Bay	Dredging and backfilling	Full-scale	2001	34,000
GM Central Foundry, St. Lawrence River, NY	PCBs	River	Dredging (backfilling in one area)	Full-scale	1995	14,000
Grasse River, NY (non-time-critical removal action)	PCBs	River	Dredging	Pilot	1995	3,000
Grasse River, NY remedial options pilot study (ROPS)	PCBs	River	Dredging and backfilling	Pilot	2005	30,000
Lake Jarnsjon, Sweden	PCBs	Lake	Dredging	Full-scale	1993-1994	196,000[g]

(Continued on next page)

TABLE 3-1 Continued

Project	Primary Chemicals of Concern	Water-Body Type	Remedy Type	Scale of Effort	Dates of Dredging	Volume of Dredged Sediment (cy)
Manistique Harbor, MI	PCBs	Harbor	Dredging	Full-scale	1995-2000	190,000[b]
Reynolds Metals, St. Lawrence River, NY	PCBs, PAHs, total dibenzofuran	River	Dredging (backfilling in one area)	Full-scale	2001	86,000
Marathon Battery, Hudson River, Cold Spring, NY	Cd	River	Dredging	Full-scale	1993-1995	71,000
New Bedford Harbor, MA (hot spot)	PCBs	Estuary	Dredging	Full-scale (hot spot removal)	1994-1995	14,000
United Heckathorn, Richmond, CA	DDT, dieldrin	Estuarine channel	Dredging and backfilling	Full-scale	1996-1997	110,000

[a]In the review and collection of data on dredging sites it was determined that various sources will often provide inconsistent information. For example, the committee found several different estimates for the volume of sediment removed from the Black River: 49,700 cy (Zarull et al. 1999; Cleland et al. 2000); 45,000 cy (Malcolm Pirnie and TAMS Consultants, Inc. 2004); 50,000 cy (EPA 2000); 60,000 cy (GE 2000; GE 2004; Fox R. Group 1999).

[b]From Dalton, Olmsted & Fuglevand, Inc. 2006.

c15,000 cy removed with upland-based equipment and 404,000 removed with marine-based equipment.

dFrom EPA summary (EPA 2006 [Commencement Bay—Sitcum Waterway, April 26, 2006]): "Approximately 396,000 cy of sediment were dredged from Area A using hydraulic and clamshell dredges and approximately 32,300 cy were removed from Area B using a small hydraulic dredge."

eFrom EPA summary (EPA 2006 [Puget Sound Naval Shipyard, May 15, 2006]).

fFrom EPA summary (EPA 2006 [Harbor Island Lockheed Shipyard Sediment Operable Unit, May 11, 2006]): "Remediation dates: These dates refer to dredging activities, not other remedial activities such as capping: November 22, 2003 to March 10, 2004 and from October 22, 2004 to November 22, 2004."

gBremle et al. 1998.

hEPA 2006 (Manistique River and Harbor Site, May 10, 2006).

Sources: Data are also unpublished data from EPA, NAS Completed Dredging Summary, March 22, 2006; Malcolm Pirnie, Inc., and TAMS Consultants, Inc. 2004.

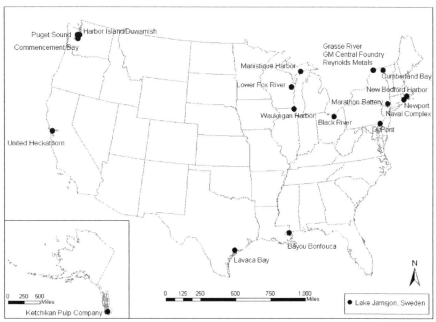

FIGURE 3-1 Locations of environmental-dredging projects selected for evaluation. Several locations comprise more than one site, including Commencement Bay, Grasse River, Lower Fox River, and Harbor Island.

River Operable Unit 1 and the Grasse River Remedial Options Pilot Study (Connolly et al. 2006; Fox et al. 2006). Those two projects were selected for evaluation because extensive monitoring data on them were made available to the committee. The committee selected a pilot dredging study in Lavaca Bay, TX, and a hot spot removal in New Bedford Harbor, MA, because they were the subject of extensive monitoring efforts. The Reynolds Metals Superfund site in the St. Lawrence River was selected because it has undergone monitoring for several years after initial dredging (EPA 2005b).

Of the 26 dredging projects, five have been identified by EPA as contaminated sediment megasites, that is, sites where the dredging portion of the remedy will cost at least $50 million (see "Sediment Contamination at Superfund Sites" in Chapter 2).

APPROACH TO EVALUATING DREDGING PROJECTS

The committee evaluated the effectiveness of the 26 dredging projects, where data permitted such an evaluation, by determining whether cleanup levels and remedial action objectives were achieved. The committee used lessons learned from the projects to identify means for improving future decisions about remediation at Superfund megasites. The lessons learned involved:

- Factors that contributed to the success of dredging operations.
- Factors that adversely affected dredging operations.
- Methods for improving the monitoring of dredging effectiveness.

The committee did not attempt to substitute its own judgment about what remedial action objectives and cleanup levels should be, including the site-specific risk modeling on which they were based, but simply tried to determine whether the stated goals were achieved and why or why not. The committee chose to review sites where dredging was the only remedy (or at least the main remedy) to evaluate the effectiveness of only dredging. However, the committee acknowledges that the effects of dredging may be difficult or impossible to distinguish from other ongoing processes including additional source control and natural recovery.

The committee recognizes the importance of quantifying the net risk reduction associated with dredging, which involves consideration of possible *increases* in risk to the spectrum of human and environmental receptors that might be caused by dredging and the treatment, storage, transport, and disposal of dredged sediment as well as risk reduction from removal of contaminated sediments from aquatic systems. The need for considering net risk reduction and incorporating the analysis into decision-making at Superfund sites has been noted in previous National Research Council reports (NRC 2001; NRC 2005), recognized by EPA (EPA 2005a), and examined in the scientific literature (Wenning et al. 2006). However, such analysis was not possible in this case. It was difficult (in some cases impossible) to obtain requisite data on the dredging component of each project, much less some of the other measures relevant to a net risk reduction analysis (for example, number of acci-

dents during transportation of dredged material) that are often not compiled. These limitations combined with time and resource constraints
prevented the committee from conducting net risk reduction evaluations
for each dredging project. As a result, the committee chose to focus its
time and resources on the dredging component of projects rather than
treatment, storage, transport, and disposal.

Methods Used to Evaluate Dredging Projects

Each project review began with the identification of the remedial
action objectives and cleanup levels for the project. That was followed by
review of project data to judge whether the remedial action objectives
and cleanup levels had been met. Those data could include chemical
concentrations in sediment, surface water, the biota, and other environmental media; indicators of exposure (such as bioaccumulation testing
and human biomonitoring data); and measures of risk (such as toxicity
testing). The measures have inherent natural heterogeneity and are subject to uncertainty. When, despite the variability and uncertainty, sufficient data were available to compare pre-dredging and post-dredging
conditions, the committee attempted to answer the effectiveness questions listed in Box 3-4. Ideally, the evaluation should be performed in the
context of a comparison of all available remedial alternatives (EPA
2005a; Bridges et al. 2006), but the committee reviewed only dredging
projects in accordance with the statement of task.

Cleanup levels and remedial action objectives are not always risk-
based, and some projects have no cleanup levels or other quantitative
means to judge effectiveness at all. Therefore, the committee looked beyond remedial action objectives and cleanup levels to identify dredging
successes and failures in reducing exposure or risk, as well as the site,
the remedy, or the contaminant conditions that led to the successes and
failures. In its evaluation of project data, the committee distinguished
between changes in environmental-media concentration, exposure, and
risk.

The committee evaluated whether baseline assessment data were
available to define conditions before dredging for comparison with
monitoring data collected during and after dredging. Optimally, those

---

BOX 3-4 Measures of Sediment-Remedy Effectiveness
in the Superfund Program

Dredging effectiveness is the achievement of cleanup goals (remedial action objectives and cleanup levels that are derived from appropriate site-specific risk modeling) in the predicted time frame with a reasonable degree of confidence that the achievement will be maintained.

Interim Measures

1. Short-term remedy performance. For example, have sediment cleanup levels been achieved after dredging?
2. Long-term remedy performance. For example, have sediment cleanup levels been maintained for at least 5 years, and thereafter as appropriate?
3. Short-term risk reduction. For example, have remedial action objectives been achieved? Do data demonstrate or at least suggest a reduction in fish tissue concentrations, a decrease in benthic toxicity, or an increase in species diversity or other community indexes after 5 years?

Key Measure

4. Long-term risk reduction. For example, have remedial action objectives been maintained for at least 5 years, and thereafter as appropriate?
5. Has the predicted magnitude and timing of risk reduction been achieved or are they likely to be achieved?

Source: Adapted from EPA 2005a.

---

assessment data would suffice to quantify exposures and risks and allow comparison with during-dredging and post-dredging monitoring data. Human and ecologic exposure and risk might increase during dredging, and these increases should also be weighed against exposure and risk reductions following dredging. Monitoring data and information collected during dredging were reviewed to identify changes in human and ecologic exposure and risk that occurred during this period. Dredging of contaminated sediment disrupts the bottom substrate, thereby destroying the existing benthic community, and can increase exposure of humans and the biota, depending on the degree of resuspension, residual generation, and release of sediment-bound, dissolved, or airborne contamination. The committee also reviewed monitoring data and information collected after dredging to identify changes in human and ecologic exposure and risk that resulted from dredging and to evaluate whether the changes should be expected to be maintained despite extreme weather conditions and human activities.

The committee used lessons learned from individual dredging project evaluations to inform its deliberations about how to improve future remediation decisions at Superfund megasites. Specifically, the committee sought to define site conditions and project design implementation factors that affect dredging success and to use this information in recommending improved management and monitoring to facilitate scientifically based and timely decision-making.

REFERENCES

Baker, J.C., W.F. Bohlen, R. Bopp, B. Brownawell, T.K. Collier, K.J. Farley, W.R. Geyer, and R. Nairn. 2001. PCBs in the Upper Hudson River: The Science behind the Dredging Controversy. A White Paper Prepared for the Hudson River Foundation, New York, NY. October 25, 2001 [online]. Available: http://www.seagrant.sunysb.edu/HEP/archive/hrfpcb102901.pdf [accessed Aug. 16, 2006].

BBL (Blasland, Bouck & Lee, Inc). 2000. Allied Paper, Inc./Portage Creek/Kalamazoo River Superfund Site RI/FS Feasibility Study Report—Phase I Allied Paper, Inc., Appendix D. Site Profiles of Sediment Dredging Projects. Draft for State and Federal Review. October 2000 [online]. Available: http://www.deq.state.mi.us/documents/deq-erd-kzoo-FS-apend-d.pdf [accessed Jan. 8, 2006].

Bremle, G., L. Okla, and P. Larsson. 1998. PCB in water and sediment of a lake after remediation of contaminated sediment. Ambio 27(5):398-403.

Bridges, T.S., S.E. Apitz, L. Evison, K. Keckler, M. Logan, S. Nadeau, and R.J. Wenning. 2006. Risk-based decision making to manage contaminated sediments. Integr. Environ. Assess. Manage. 2(1): 51-58.

Cleland, J. 2000. Results of Contaminated Sediment Cleanups Relevant to the Hudson River: An Update to Scenic Hudson's Report "Advances in Dredging Contaminated Sediment." Prepared for Scenic Hudson, Poughkeepsie, NY. October 2000 [online]. Available: http://www.scenichudson.org/pcbs/pcb_dredge.pdf [accessed Aug. 16, 2006].

Connolly, J., V. Chang, and L. McShea. 2006. Grasse River Remedial Options Pilot Study (ROPS) Findings from Dredging Activities. Presentation at the

Second Meeting on Sediment Dredging at Superfund Megasites, June 7, 2006, Irvine, CA.

Dalton, Olmsted & Fuglevand, Inc. 2006. Remediation Action Construction Report, Part 1: Head of Hylebos Waterway Problem Area Commencement Bay Nearshore/Tideflats Superfund Site Tacoma, Washington, Review Draft, July 21, 2006. Prepared for Head of Hylebos Cleanup Group, Arkema, Inc, General Metals of Tacoma, Inc, by Dalton, Olmsted & Fuglevand, Inc., Kirkland, WA. July 21, 2006.

Ells, S. 2006. Summary of Available Information on Superfund Sediment Sites and Dredging Sites. Presentation at the First Meeting on Sediment Dredging at Superfund Megasites, March 21-22, 2006, Washington, DC.

EPA (U.S. Environmental Protection Agency). 1998. Realizing Remediation: A Summary of Contaminated Sediment Remediation Activities in the Great Lakes Basin. U.S. Environmental Protection Agency, Great Lakes National Program Office, Chicago, IL. March 1998 [online]. Available: http://www.epa.gov/glnpo/sediment/realizing/realcover.html [accessed Aug. 16, 2006].

EPA (U.S. Environmental Protection Agency). 2000. Realizing Remediation II: A Summary of Contaminated Sediment Remediation Activities at Great Lakes Areas of Concern. U.S. Environmental Protection Agency, Great Lakes National Program Office, Chicago, IL. July 2000 [online]. Available: http://www.epa.gov/glnpo/sediment/realizing2/index.html [accessed Aug. 16, 2006].

EPA (U.S. Environmental Protection Agency). 2005a. Contaminated Sediment Remediation Guidance for Hazardous Waste Sites. EPA-540-R-05-012. OSWER 9355.0-85. Office of Solid Waste and Emergency Response, U.S. Environmental Protection Agency, December 2005 [online]. Available: http://www.epa.gov/superfund/resources/sediment/guidance.htm [accessed Jan. 8, 2007].

EPA (U.S. Environmental Protection Agency). 2005b. Five-Year Review Report Reynolds Metals Company Site St. Lawrence County Town of Massena, New York. U.S. Environmental Protection Agency, Region 2, New York. April 2006 [online]. Available: http://www.epa.gov/superfund/sites/fiveyear/f06-02018.pdf [accessed Jan. 8, 2007].

EPA (U.S. Environmental Protection Agency). 2006. Case Study Data and Information. U.S. Environmental Protection Agency.

EPA and Environment Canada (U.S. Environmental Protection Agency and Environment Canada). 2005. Great Lakes Binational Toxics Strategy: 2005 Annual Progress Report. U.S. Environmental Protection Agency, Chicago, IL, and Environment Canada, Environmental Protection Branch, Downsview, ON [online]. Available: http://binational.net/bns/2005/2005-GLBTS-English-web.pdf [accessed Sept. 12, 2006].

Fox River Group. 1999. Effectiveness of Sediment Removal. September 27, 1999 (Appendix C in GE 2000)

Fox, R., M. Jury, and J. Kern. 2006. Overview of Lower Fox River OU1 Dredge Residuals. Presentation at the Second Meeting on Sediment Dredging at Superfund Megasites, June 7, 2006, Irvine, CA.

GE (General Electric Company). 2000. Environmental Dredging: An Evaluation of Its Effectiveness in Controlling Risks. General Electric Company, Blasland, Bouck & Lee, Inc., and Applied Environmental Management, Inc., Albany, New York. August 2000 [online]. Available: http://www.ge.com/files/usa/en/commitment/ehs/hudson/DREDGE.pdf [accessed Jan. 8, 2006].

GE (General Electric Company). 2004. Major Contaminated Sediments Sites (MCSS) Database, Release 5.0. General Electric Company, Applied Environmental Management, Inc., and Blasland, Bouck & Lee, Inc. September 17, 2004 [online]. Available: http://www.ge.com/en/citizenship/ehs/remedial/hudson/mcss/database.htm [accessed Jan. 8, 2006].

Great Lakes Dredging Team. 2006. Case Study Series [online]. Available: http://www.glc.org/dredging/ [accessed Sept. 12, 2006].

Hahnenberg, J. 1999. Long-term benefits of environmental dredging outweigh short-term impacts. Engineering News Record (March 22/29):E15-E16.

Malcolm Pirnie, Inc, and TAMS Consultants, Inc. 2004. Final Engineering Performance Standards Hudson River PCBs Superfund Site, Appendix: Case Studies of Environmental Dredging Projects. Prepared for U.S. Army Corps of Engineers, Kansas City District, by Malcolm Pirnie, Inc., White Plains, NY, and TAMS Consultants, Inc., Bloomfield, New Jersey. April 2004 [online]. Available: http://www.epa.gov/hudson/eng_perf/FP5001.pdf [accessed Aug. 16, 2006].

Nadeau, S. 2006. SMWG Review and Analysis of Selected Sediment Dredging Projects. Presentation at the Second Meeting on Sediment Dredging at Superfund Megasites, June 7, 2006, Irvine, CA.

NRC (National Research Council). 2001. A Risk-Management Strategy for PCB-Contaminated Sediments. Washington, DC: National Academy Press.

NRC (National Research Council). 2005. Superfund and Mining Megasites: Lessons from the Coeur d'Alene River Basin. Washington, DC: The National Academies.

Pastor, S. 1999. Dredging: Long-term benefits outweigh short-term impacts. Fox River Current 2(4):7-8 [online]. Available: http://www.epa.gov/Region5/sites/foxriver/current/foxcurrent199909.pdf [accessed Sept. 12, 2006].

Scenic Hudson. 2000. Accomplishments at Contaminated Sediment Cleanup Sites Relevant to the Hudson River. Scenic Hudson, Poughkeepsie, NY. September 2000.

Thibodeaux, L.J., and K.T. Duckworth. 2001. The effectiveness of environmental dredging: A study of three sites. Remediation J. 11(3):5-34.

USACE (U.S. Army Corps of Engineers). 2006. Case Studies and Projects [online]. Available: http://el.erdc.usace.army.mil/dots/ccs/case.html [accessed Sept. 12, 2006].

Wenning, R.J., M. Sorensen, and V.S. Magar. 2006. Importance of implementation and residual risk analyses in sediment remediation. Integr. Environ. Assess. Manage. 2(1):59-65.

Zarull, M.A., J.H. Hartig, and L. Maynard. 1999. Ecological Benefits of Contaminated Sediment Remediation in the Great Lakes Basin. Windsor, Ontario: International Joint Commission [online]. Available: http://www.ijc.org/php/publications/html/ecolsed/ [accessed Sept. 12, 2006].

4

Evaluation of Dredging Effectiveness: What Has Experience Taught Us?

INTRODUCTION

Over the last 20 years, various contaminated sediment sites have been remediated in whole or in part through dredging. The committee examined 26 dredging projects to evaluate whether dredging was able to meet short-term and long-term goals. Short-term goals are defined as cleanup levels that can be measured during or immediately post-dredging to verify the effective implementation of the remediation. The ability to maintain cleanup levels in the long term is ideally linked to the achievement of long-term risk-based goals or remedial action objectives. Appendix C presents the various sites' cleanup levels and remedial action objectives and describes whether they were achieved at individual sites. Taken as a whole, the projects indicate what can and cannot be achieved with dredging and the conditions that favor or discourage the use of dredging.

Evidence that dredging projects led to the achievement of long-term remedial action objectives and did so within expected or projected time frames is generally lacking. It was often not possible to evaluate long-term remedy performance relative to remedial action objectives because of insufficient post-remediation data, quality, or availability or because of lack of an equivalent pre-remediation dataset. Post-remediation conditions are always influenced by long-term natural attenuation processes and ongoing sources of contaminations if they exist, so long term monitoring—over decades—may be needed to establish effectiveness; few of the sites reviewed by the committee have reached that level of maturity. Not counting the 5 pilot studies or hot-spot removal actions, about one half of the sites apparently did not achieve remedial action objectives or had inadequate monitoring to judge performance relative to remedial action objectives. Insufficient time has elapsed to judge achievement of remedial action objectives in approximately one quarter of the sites. The remaining sites apparently met remedial action objectives although the extent to which those remedial action objectives achieve long-term risk reduction may not be known.

There were often sufficient data to evaluate performance relative to cleanup levels or short-term implementation goals, but the relationship of these measures to long-term risk reduction was often not clear. An examination of Appendix C shows that many sites achieved cleanup levels; however, many were operational goals (mass removal or dredging to elevation) rather than contaminant-specific goals.

Natural processes are always modifying conditions at a site; their influence can be difficult or impossible to separate from the remedial action, particularly when control or reference sites are not monitored before and after remediation. Conditions also are often influenced by the implementation of combined remedies, such as dredging and capping, which complicate the assessment of the performance of dredging alone. Thus, the committee was unable to evaluate the effectiveness of dredging *alone* at most sites.

Experiences at the sites can nevertheless inform remedial project managers as to what may be achievable with dredging and what site and operational factors may limit dredging effectiveness or contribute to its success. Experience is especially useful in identifying factors that contribute to success or failure of dredging to meet short-term cleanup levels because monitoring has been conducted at most sites to judge performance relative to these standards. The ability to meet chemical-specific cleanup levels, however, does not in itself mean the ability to meet long-term risk-reduction targets or indicate the time frame over which any such targets might be met. This chapter discusses the lessons learned from sites where dredging was conducted and uses specific examples to

illustrate them. It also provides recommendations regarding implementation of successful remediation with dredging.

DATA AVAILABILITY AND ACCESSIBILITY

The potential utility of a review of remedial effectiveness is gov-
erned by the availability of pre-remediation and post-remediation monitoring data. That issue has three components: whether data were acquired at a site, whether the data are available for review, and whether
the data are sufficient to support conclusions about effectiveness. The
goal of acquiring data for this type of analysis appears relatively simple:
collect and evaluate pre-remediation and post-remediation monitoring
data on concentrations and effects from Superfund sites. However, obtaining this information is surprisingly difficult.[1,2]

---

[1]For example, the post-dredging sediment concentrations at the Waukegan
Harbor, IL, Superfund site were of interest. In response to the committee's re-
quest for these data, the U.S. Environmental Protection Agency (EPA) stated only
"Post-Cleanup: 1/1/2005, Ave. 2.5 ppm (Source: RPM)," with no supporting data
(EPA 2006a [OMC Waukegan Harbor Site, May 15, 2006]). To pursue sediment
data further, the committee reviewed the site's 5-year reviews (EPA 1997a, 2002).
The 1997 5-year review states that "confirmation sampling was taken at the base
of dredge to verify that contaminants levels required for this cleanup were met."
However, the sampling did not include chemical analyses of contaminant levels
in the sediment (Canonie Environmental 1996). The 2002 5-year review does not
indicate that any post-remediation monitoring of the harbor sediments had been
conducted. However, another literature search indicated that EPA contaminants
studied in harbor sediments collected in 1996, a few years after dredging (EPA
1999) and again in 2003 (ILDPH/ATSDR 2004).

[2]For example, the post-dredging sediment concentrations at the Bayou Bonfouca, LA, Superfund site were of interest. In EPA's summary to the committee
(EPA 2006a [Bayou Bonfouca Superfund site, May 12, 2006]), they indicate that
COC concentration data in sediment and biota are unavailable. The "monitoring"
section in EPA's summary refers only to a study by the Hazardous Substance
Research Center S&SW which contains an analysis of remedy performance but
not the raw monitoring data. However, the conclusion contains an excerpted
paragraph from a 2003 report on the site (EPA 2003a) that indicates that post-
dredging sediment samples were collected (a portion of the 2003 report [without
sample locations or relation to the dredging site] was provided). The locations
were later received from EPA. To pursue obtaining sediment data further, the

The amount, frequency, and type of data collected at dredging sites are highly variable. Some of the earlier sites had very little post-dredging monitoring. For example, Bayou Bonfouca, LA, and Outboard Marine Corporation, IL, did not sample sediment concentrations immediately after dredging (see footnotes 1 and 2). Marathon Battery, NY, is an exception in that sediment and biota concentrations were collected and bioaccumulation was tested, but obtaining monitoring data proved difficult, requiring several iterations, and ultimately the committee could not access the full range of reports. At some more recent sites, dredging is supported and guided by chemical confirmation sampling, and the resulting data are accessible. For example, the U.S. Environmental Protection Agency (EPA) provided the committee concentration and location data on the recently completed dredging in the Grasse River, NY, and Hylebos Waterway in Commencement Bay, WA, including date, location coordinates, and chemical concentration information. Information on the operations and sampling and the monitoring results at pilot projects (such as conducted at the Grasse River, NY; Fox River, WI; and Lavaca Bay, TX, sites) are often well documented, as would be expected from studies specifically intended to document remedial effectiveness on a smaller scale.

Data and reports from remediation sites are often held by various entities (including EPA, consultants, states, and responsible parties), and

---

site's 5-year reviews (EPA 1996a, 2006b; CH2M Hill 2001) were examined. The 2001 5-year review (the first to address the dredging activity) states that "no monitoring of the water level or quality conditions in the bayou are currently conducted—and no water quality data has been collected in the bayou adjacent to the site since the end of the source removal remedial action in 1995." However, it also states that "the swimming and sediment contact advisory remains in effect based on the sediment samples collected [by the State of Louisiana] in 1997." The 5-year review (CH2M Hill 2001) does not provide these sampling data or indicate locations or average concentrations of the samples. Later efforts to obtain the data through EPA were not successful. The 2006 5-year review indicates that sediment samples were taken after Hurricane Katrina in 2005. The post-hurricane sampling report was not provided to the committee, but the data (without sample locations) were provided in the 2006 5-year review. The post-hurricane report (CH2MHILL 2006), with sample locations, was later requested and acquired from EPA.

this complicated the compilation of information. Committee requests for data on sites were sent to Superfund Headquarters, which did not have them and forwarded the requests to project managers in the EPA region who were responsible for a particular site (although these managers may or may not have been in that position during the remediation). The data might not be held by the EPA region, but instead may reside with the contractors that performed the work or the responsible parties that funded the work. Some data at sites where remedial actions had been completed are archived or not readily retrievable. Thus, even when information was available to EPA it might have been inaccessible. In some cases, reports containing monitoring data and interpretations were held by the responsible parties but EPA wished not to have them released because sensitive negotiations were under way.

When reports and data were available, they may have been reproduced only on paper although they were originally produced electronically. Such conditions severely limit distribution and faithful replication of information, because many site documents rely on large-scale maps in color. The ability to access reports and data via public Web sites was generally extremely limited, but there were exceptions. The mid-Atlantic EPA Region 3 has each site's administrative records on line, and this permits the public and researchers to access site files electronically (although typically these files are in a scanned, nonnative format).[3] Public information is available on all Superfund sites via the CERCLIS database, which frequently contains a site's record of decision and 5-year reviews, if available. It was presumed that a site's 5-year review would contain explicit statements of the sampling that had been conducted and provide, at a minimum, concentration and location data on sampling, but the committee was surprised to see that that was not necessarily the case.[1,2]

Comments regarding the ready public accessibility of electronic data may seem trivial. However, pre-remediation and post-remediation information is the end result of massive planning, implementation, and

---

[3]The administrative records contain much information that is ancillary to understanding site conditions (for example, e-mails, records of phone discussions, submitted comments, and written communications among states, agencies, and responsible parties). Those materials are important for maintaining a transparent decision-making process, but the primary data reports and summaries are most important for reviewing remedial effectiveness.

data collection efforts that typically have involved large expenditures of public resources (whether the expenditures result from remediation itself or from the establishment of agreements with responsible parties who conduct it). Provision of pre-remediation and post-remediation data on chemicals of concern in an accessible, intuitive manner that defines collection efforts and results is a prerequisite to reviewing and understanding the results of remedial projects.

An issue related to the availability (or lack) of complete sampling data is the need to rely on data summaries and various site reports to evaluate pre- and post-remediation results. At times the committee relied on reports and summaries that did not convey the necessary raw data to confirm summary statistics. That is because the committee did not have access to the primary sources or the resources to complete an ad-hoc reassembly and evaluation of all the information. A note of caution relevant to this and other studies that summarize site information is that data on concentrations and effects should be collected consistently over time (for example, from the same locations, media, depth interval, and developed with similar techniques and protocols) to be most useful. Summary statistics may not be derived from similar datasets and reports may have incomplete annotation on sample location (for example, the relation of the samples to the dredging footprint), sampling protocols, and chemical analyses. Over time, analytical methods, contractors, sampling locations, and sampling methods can change. These changes complicate pre- and post-remediation comparisons. When possible, the committee provides information on these issues.

DREDGING EFFECTIVENESS

Dredging to Remove Contaminant Mass

The direct effect of dredging is the removal of sediment and its associated contaminant mass. Experience at a variety of sites has shown that dredging is effective at removing contaminant mass. Where sediments are subject to scour by storm or other high-flow events, buried contaminated sediment may be the source of future exposure and risk. In such cases, mass removal may result in risk reduction because the future exposure and transport of sediment have been thwarted (see Chapter 2 for additional discussion).

For example, a demonstration dredging project was conducted to remove a deposit (Deposit N) contaminated with polychlorinated biphenyls (PCBs) in a high-velocity reach of the Lower Fox River, WI, in 1998 and 1999. PCB contamination in sediments of the river is the result of historical wastewater discharge from the manufacture and recycling of carbonless copy paper incorporating Aroclor 1242. The objective of the Deposit N demonstration was to remove contaminated sediment and leave no more than 3-6 in. of residual material in place while minimizing resuspension and offsite loss of sediment. Dredging to target elevations in 1998 and 1999 resulted in the removal of 112 kg of PCBs, or 78% of the pre-dredging inventory (Foth and Van Dyke 2000). Mass removal may have been an appropriate cleanup objective if there was the potential for future mobilization and transport of the PCBs.

Simple mass removal, however, may not reduce risk. For example, the non-time-critical removal action conducted in 1995 in the Grasse River in Massena, NY, had the objective of removing much of the PCB mass that was in the vicinity of an outfall. PCBs had been in use at the adjacent Alcoa facility and were introduced into the river through the outfall and from other sources. It was estimated that this localized removal of about 2,500 cy removed 27% of the PCB mass from the entire study area, consisting of several miles of river (BBL 1995). Despite removal of as much as 98% of the targeted contaminant mass (Thibodeaux and Duckworth 1999), no measurable reduction in water-column or fish concentrations of PCBs was noted. Site characterization and assessment efforts have led to the conclusion that water-column PCB concentrations are related, at least during low-flow periods, to surficial sediment concentrations of PCBs throughout the river and that removal of buried mass does not have a major influence on water-column concentrations (Ortiz et al. 2004). The removal may still have been warranted to avoid potential scouring during high-flow conditions, but risk reduction was not achieved during base flow conditions.

Dredging to Reduce Risk

A more complete assessment of dredging effectiveness would include evaluation of long-term risk reduction in addition to mass removal or performance relative to cleanup levels. Although few sites have sufficiently complete datasets, dredging has apparently resulted in risk reduction in some cases, including at some sites with long-term datasets.

Lake Jarnsjon, Sweden

The Lake Jarnsjon site was remediated in 1993 and 1994. The pre-dredging surface sediment (0-40 cm) PCB concentrations had a geometric mean of 5.0 mg/kg (n=12; range 0.4 to 30.7 mg/kg) in 1990. Following remediation, the surface sediment (0-20 cm) concentrations in 1994 were significantly reduced and had a geometric mean of 0.060 mg/kg (n=54; range 0.01 to 0.85 mg/kg) (Bremle et al 1998a). Out of 54 defined subareas, one exceeded 0.5 mg/kg dry weight (set as the highest acceptable level to be left in the sediment); 20% of the sediment areas had PCB levels higher than 0.2 mg/kg dry weight, the remediation objective was set at 25% (Bremle et al 1998b).

Fish-tissue PCB concentrations declined after remediation although post-dredging monitoring did not take place until 2 years after dredging. Concentrations did not, however, decrease to those upstream of the contaminated area. In their report, Bremle and Larsson (1998) compare fish concentrations in the remediated lake to fish in upstream areas and conclude that "fish from all the locations in 1996 had lower PCB concentration than in 1991 [dredging occurred in the summers of 1993 and 1994]. The most pronounced decrease was observed in the remediated lake, where levels in fish were halved. The main reason for the reduced levels was the remediation." However, the authors also state that "the reason for the decline of PCB in fish could be decreased atmospheric deposition and thus lowered loadings of PCB to the freshwater."

The comparisons in the study benefited from the use of a reference site that indicated background declines in fish-tissue concentrations; these declines have been seen elsewhere as well (Stow et al. 1995). In that regard, Bremle and Larsson (1998) state that

> ...the results show that if a remedial action is to be evaluated and the process is extended over several years, changes in background contamination must be taken into account. After a remedial action, the results need to be followed over several years to show if it has

been successful, which has not yet been the case in the present study. It also stresses the importance of using reference sites, to compare the results from the remedial area. A decrease in overall background contamination could otherwise well be interpreted as a result of the remedial action only.

Black River, Ohio

At the Black River, in Lorain, OH, sediment was contaminated with polycyclic aromatic hydrocarbons (PAHs) as a result of effluent from a steel-plant coking facility. In 1983, the coking facility closed. Dredging occurred from late 1989 to early 1990 below the Kobe Steel outfalls at river miles 2.83-3.55 (EPA 2007a). In the early 1980s, PAH compounds were detected in sediments at high concentrations, and the brown bull-head population had high rates of liver cancer and pre-cancerous lesions. Since closure of the facility and dredging, PAH concentrations in surface sediments, fish PAH residues, and neoplasm frequencies in fish have declined (Baumann 2000). As shown in Figure 4-1, a decrease in cancer at the site was noted immediately after the plant closure. An increase in cancer was also noted immediately after dredging and was probably due to the exposure of fish and their prey to higher concentrations of PAHs in sediment and water during dredging. Later sampling, however, showed decreases in cancer, suggesting that the increase during dredg-ing was a short-term phenomena. Within 5 years after remediation, the cancer incidence was lower than the pre-dredging data, presumably as a result of the dredging. However, it is unclear to what extent continued natural attenuation, as evidenced by the reduction in observed cancer after plant closure but before dredging, could have reduced cancer inci-dence in the same time frame.

Marathon Battery, New York

The ability to achieve remedial action objectives and long-term risk reduction with dredging was demonstrated at Foundry Cove of the Marathon Battery, NY, site. Foundry Cove is a small body of water adja-cent to the Hudson River about 85 km north of New York City. The

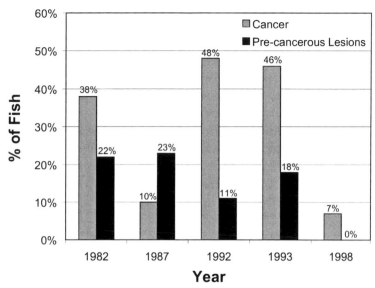

FIGURE 4-1 Prevalence of cancer and pre-cancerous lesions in Black River bull-heads before and after dredging. The primary source of contamination was shut down in 1983, and dredging occurred in 1989-1990. Source: Adapted from Baumann 2000.

Marathon Battery Company discharged cadmium, nickel, and cobalt during the manufacture of batteries through the plant's outfalls, located beneath the Cold Spring pier and in the East Foundry Cove Marsh. About 50 metric tons of nickel-cadmium waste is estimated to have been discharged from 1953 to 1971 (Levinton et al. 2006).

The site comprises six separate regions. West Foundry Cove borders the eastern shore of the Hudson River and is connected to East Foundry Cove via an opening in a railroad trestle. The most contaminated sites are East Foundry Cove Marsh (13 acres), East Foundry Cove (36 acres), East Foundry Pond (3 ac), and the Cold Spring Pier area (~5 acres) that borders the Hudson River to the north. Constitution Marsh (281 acres) is to the south and is less contaminated (see Figure 4-2). As summarized in the record of decision and summary to the committee, core sediment samples collected from East Foundry Cove during the remedial investigation ranged from 0.29 to 2700 mg/kg cadmium and had

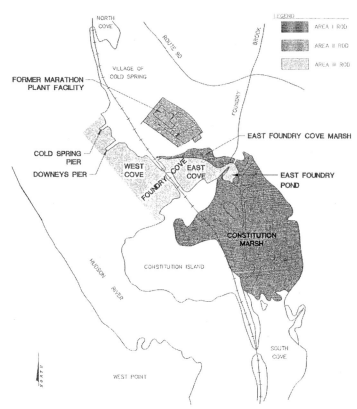

FIGURE 4-2 Map showing the Marathon Battery Superfund site on the Hudson River, NY adjacent to Cold Spring, NY. Dredging was conducted in the Cold Spring Pier area, East Foundry Cove, and East Foundry Pond. East Foundry Cove Marsh was excavated and capped; Constitution Marsh was not remediated. Source: EPA 2006a (Marathon Battery Superfund Site, May 10, 2006).

a mean of 179.3 mg/kg (median = 5.6 mg/kg).[4] Samples collected in the Pier area (a much larger area than what was actually dredged) ranged from 1.2 to 1,030 mg/kg for cadmium and had a mean of 12.6 mg/kg

---

[4]These data are for all depths. The mean for each sampled depth is 439.4 mg/kg (0-10 cm), 50.5 mg/kg (10-25 cm), and 2.1 mg/kg (25-50 cm). This sampling was apparently conducted in 1984 by Acres in support of the remedial investigation (EPA 1989b). These summary statistics from EPA do not include other sampling data collected in 1989 (USACE 1992).

(median = 3.9 mg/kg)[5] (EPA 1989a; 2006a [Marathon Battery Superfund Site, May 10, 2006]). A human health risk assessment that considered fish and crab consumption and exposure to suspended sediment in water concluded that achievement of a sediment cadmium concentration of 220 mg/kg would be protective. Sediment bioassays indicated that 10-255 mg/kg would be protective of aquatic life. A cleanup level of 10 mg/kg was set and believed to be achievable by removing the top 1 ft of sediment (EPA 1989a). About 80,000 cy of sediment was dredged from the contaminated areas of East Foundry Cove, East Foundry Cove Pond, and the Cold Spring Pier area from 1993 to 1995. In contrast, East Foundry Cove Marsh was dry excavated to a 100 mg/kg limit and then capped with a bentonite and geotextile blanket followed by 1 ft of sandy marsh planting material. No active remediation was implemented in the Constitution Marsh[6] or the West Foundry Cove (EPA 1995; 2006a [Marathon Battery Superfund Site, May 10, 2006]).

Post-dredging verification sampling was conducted to establish whether cleanup levels had been met. EPA states "In the Hudson River and East Foundry Cove, an average of 10 mg/kg cadmium remained, which was consistent with the ROD requirement that at least one foot of sediment and 95% of the contamination be removed" (EPA 1998a). The record of decision also required long-term monitoring at the site for thirty years after completion of the remedial action (AGC 2001). Figure 4-3 presents median cadmium concentrations[7] at the East Foundry Cove portion of the site before and after remediation from this monitoring program. These data indicate that surficial sediment concentrations were reduced as a result of dredging, and the concentrations have not re-

---

[5]These data are for all depths. The ROD describes these data as being from 85 locations covering 465 acres with cores down to a depth of 137 cm. This is a greater area than the approximate 5 acres that was dredged.

[6]"Although cadmium-contaminated sediment hot spots were identified in Constitution Marsh, to remediate these sediments would have had a significant adverse impact on the marsh's sensitive ecosystem. In addition, the cadmium-contaminated sediments would eventually be covered with clean sediments following the remediation of the cadmium contaminated sediments in East Foundry Cove marsh. Therefore, long-term monitoring was selected for Constitution Marsh" (EPA 1995).

[7]For information on the distribution of data, see Figure 4-4.

EFC - Median Sediment Cd Concentrations

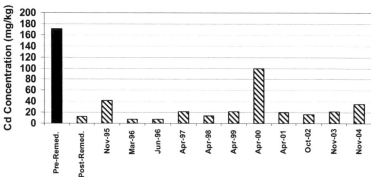

FIGURE 4-3 Median sediment cadmium concentrations, East Foundry Cove, Marathon Battery Site. Dredging occurred in the fall of 1993 and between April 1994 and February 1995. Data are the median of cadmium concentrations from five locations within the dredged area of East Foundry Cove. Pre-remediation data are from "samples obtained by Malcolm Pirnie and others prior to the remedial action. These are the reported data closest to the present LTM [long-term monitoring] sampling locations."[8] Post-remediation data (presumably confirmation sampling immediately following dredging) are the "average value of either the two closest post-remediation sample node locations or the analytical results of the various testing agencies for the same node location." The remaining data (with dates specified) are from the long-term monitoring (EPA 2006a [Marathon Battery Superfund Site, May 10, 2006]; the origin of the data table from which this figure was constructed was not noted in the summary from EPA).

turned to pre-dredging levels. Note that although cleanup levels were achieved immediately after dredging, median concentrations have since fluctuated above and below the 10 mg/kg cleanup level. Figure 4-4 presents the frequency of occurrence of sediment cadmium concentrations in the period 1995-2000. The figure shows a reduction in surficial concentration and also shows that many individual sediment samples show concentrations greater than 100 mg/kg ($log_{10} = 2$). Concentrations in some sediment samples are indistinguishable from the original sediment concentration distribution. Independent sampling and analysis by Mackie et

---

[8]It is unclear exactly when these data were collected and to what sediment depth. According to the ROD, East Foundry Cove sediment samples were collected by Acres in 1985, Ebasco in 1988, and Malcolm Pirnie in 1989.

al. (2007) had similar results. Their 2005 sampling of the dredged area of East Foundry Cove showed a median cadmium concentration of 39.2 mg/kg in the top 5 cm of sediment (16 samples ranging from 2.4 to 230.4 mg/kg, mean 59.7, SE 16.8).

The significance of occasional high residual concentrations after dredging can be evaluated by examination of ecologic exposure before and after dredging. Figure 4-5 is a summary of long-term monitoring data collected for 5 years post-dredging (AGC 2001) and shows the ratio of pre-remediation to post-remediation tissue concentrations in various plants and animals. The data most relevant to the dredging (which occurred only in East Foundry Cove, East Foundry Pond, and the Cold Spring Pier area) are the water chestnut in East Foundry Cove, which show improvement and the benthic invertebrates in East Foundry Cove, which show an increase after remediation and then a decrease to pre-remediation levels after 5 years.[9] Bioaccumulation studies (using in-situ

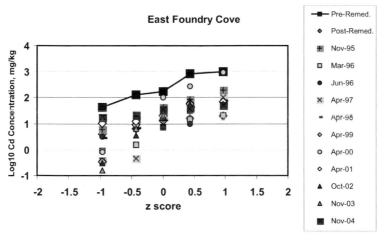

FIGURE 4-4 Log probability plot of sediment cadmium concentrations, East Foundry Cove, Marathon Battery. The z score indicates the distance from the mean in standard deviations; z scores greater than 0 exceed the mean. The source of these data is described in Figure 4-3.

---

[9]Because birds are highly mobile, contaminant exposures to wood ducks, swallows, and marsh wrens cannot be directly linked to the dredged areas. Also, Constitution Marsh did not undergo active remediation; East Foundry Cove Marsh was excavated and capped.

enclosures) were also conducted at the site. Data from East Foundry Cove and the Cold Spring Pier area generally indicate declines in accumulated cadmium body burdens compared with pre-remediation values (AGC 2001; EPA 2003b).

The data suggest that in at least some cases dredging can achieve and maintain reductions in sediment concentrations and body burdens of contaminants although occasional measurements of elevated concentrations complicate the interpretation of the results. In addition, there may be short-term increases in body burdens in species directly affected by the remediation (such as the benthic organism data). The fact that individual sediment samples exhibited elevated concentrations emphasizes that evaluation of the performance of any remedy requires adequate monitoring of key indicators before and after remediation to fully characterize the distribution of concentrations so decisions are not driven by low probability events.

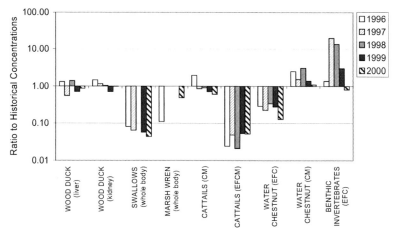

FIGURE 4-5 Ratio of cadmium body burdens in various animals and plants after dredging to pre-dredging levels for the period 1996-2000, Marathon Battery Site, NY. The data are the mean value except for the benthic invertebrates for which only one sample was collected. (The benthic invertebrate samples were "a combination of oligochaete worms and chironomid midge larvae, with a minimum aggregate weight of one gram.") The number of samples in each mean varies. CM = Constitution Marsh; EFCM = East Foundry Cove Marsh; EFC = East Foundry Cove. Note that only the water chestnuts (EFC) and the benthic invertebrates (EFC) were from locations where dredging alone was conducted. Source: Data from AGC 2001.

Summary

Dredging remains one of the few options available for the remediation of contaminated sediments and should be considered, along with other options, for managing the risks that they pose. The need to remove contaminated sediments can be particularly acute at sites where navigational channels need to be maintained or where buried contaminated sediment deposits are likely to be subjected to erosion and transport from high flows or changes in hydrologic conditions. As shown at the Grasse River and other sites, dredging can achieve removal of sediments and much of the contaminants they contain. Mass removal alone, however, may not achieve risk-based goals, which should be the basis for remedy selection.

The results at the Marathon Battery and other sites outlined above show that under some conditions dredging can achieve cleanup levels and aid recovery of biota at contaminated sediment sites. As indicated previously, it is often difficult to evaluate the effectiveness of dredging alone as a result of reductions in ongoing sources of contamination (for example, outfalls and atmospheric deposition), the use of combination remedies, and natural burial of existing contamination. As illustrated in the examples above, measurement of time trends in sediment and water contaminant concentrations prior to and after dredging can help identify changes due to dredging as well as evaluate the risks of not dredging. Reference sites are also useful in contrasting the contaminant dynamics and risk reduction due to dredging with that caused by these other processes. To assist in evaluating the performance of dredging, it is important to monitor a range of dredging performance metrics linked to risk reduction, as was done at Marathon Battery.

FACTORS AFFECTING DREDGING EFFECTIVENESS

A variety of factors and site characteristics influence dredging effectiveness and can limit the ability to achieve cleanup levels and remedial action objectives. However, it is generally not possible to definitively identify the specific conditions or factors that determined success or failure in a particular project. The committee strived to identify the condition or conditions that appeared to significantly contribute to the project's outcome and summarized those herein.

Foremost among the factors that influence dredging effectiveness are whether a site has been adequately characterized and whether ongoing sources of contamination have been controlled. Site characterization and source control require a firm understanding of the nature and distribution of the contamination, any potential sources that contribute to the contamination burden in the watershed, and the processes influencing site risks and their attenuation. A strong understanding of the extent of contaminants in place and the contribution of outside sources is essential to developing an effective conceptual site model and remedial plan to eliminate or lessen contaminant exposures and risk. The influence of source control and site characterization on remedy effectiveness will be illustrated through experience at particular sites in later sections. These factors, however, influence the success of all sediment remediation efforts, not just dredging.

Destruction of the benthic community and removal of habitat is unavoidable with all dredging projects and represents an immediate negative effect to the existing benthic community. This effect also occurs with other active remedial efforts such as capping. As such, the ecologic benefit of the current habitat needs to be an important part of the decisions in determining whether or not to dredge. In a net risk reduction framework, the habitat destruction will be compared to the benefits of removing contaminated sediment, bearing in mind the post-dredging substrate's desirability as a habitat (or the substrate created following backfilling).[10] Recovery after disturbance is typically relatively rapid with estimates of benthic recovery rates ranging from several months to several years (Qian et al. 2003; USACE 2005). Immediately after destruction of the habitat, hardy, opportunistic organisms such as polychaetes (oligochaetes in freshwater) and small bivalves colonize surficial sediments. Subsequently the population increases in diversity and abun-

---

[10]As described by EPA (2005a): "While a project may be designed to minimize habitat loss, or even enhance habitat, sediment removal and disposal do alter the environment. It is important to determine whether the loss of a contaminated habitat is a greater impact than the benefit of providing a new, modified but less contaminated habitat."

dance. Recovery occurs when the site returns to pre-disturbance conditions or does not differ significantly from a reference area.

In contrast to the above factors which affect many or all remedial approaches, dredging projects are specifically influenced by two additional processes that weigh heavily on the effectiveness of dredging:

- Resuspension of sediments and release of contamination during dredging.
- Generation of residual contamination giving rise to potential long term exposure after dredging.

As described in Chapter 2, resuspension, release, and residuals occur to some extent with all dredging projects. *Resuspension* refers to sediment that is disturbed during dredging and transported out of the dredging area. Exposure to resuspended sediments is generally transitory and ends soon after the completion of dredging. *Residuals* are the contaminated sediments exposed after conclusion of the dredging and can lead to longer term exposure and risks to organisms. *Release* of contaminants to the liquid phase (for example, solubilized PCBs) can occur from both resuspended and residual sediments. In the case of strongly sorbing contaminants, it is often assumed that the fraction of sediment resuspended corresponds to the fraction of contamination released and transported down current. Sediment resuspension and contaminant release, however, may not be closely related if there are large dissolved or separate phase releases or if release from residuals is substantial.

Figure 4-6 summarizes the amounts of resuspension and residual contamination that have been observed in a variety of dredging projects (Patmont 2006). The sediment resuspension data points are the fraction of sediment resuspended during dredging. (The fraction is the mass of suspended solids measured at some distance downstream of the dredge [typically less than 100 ft] divided by the mass dredged on a dry weight basis.) These data are from a variety of sources including consultant reports and the open literature. As such, sampling methods and approaches used to estimate these fractions can vary depending on the objectives of the study, the nature of the project, and site conditions. The residual data in the figure are the fraction of contaminant mass (not sediment mass) remaining post-dredging compared to the estimated

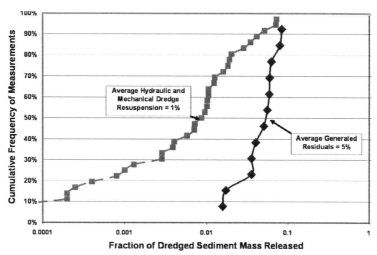

FIGURE 4-6 Sediment or contaminant mass resuspended or left as a residual. Resuspension data (squares, left) are from a variety of projects gleaned from the literature. Residuals data (diamonds, right) are from dredging at 12 projects (average refers to the median in the figures). Source: Patmont 2006. Reprinted with permission; copyright 2006, Anchor Environmental LLC.

contaminant mass removed in the last production dredging cut at 11 sites (residual fractions are determined based on a chemical mass-balance approach [see Patmont and Palermo 2007]). The projects span a range of physical settings, operating conditions, and data collection methods. Despite variations among sources of the resuspension and residuals data, the distribution shown is a useful compendium of existing data and likely to indicate at least the magnitude of expected residuals and resuspension rates.

The figure suggests that about half the dredging projects have resulted in resuspension that amounts to 1% or less of the mass of sediments dredged.[11] About half the dredging projects have resulted in a residual contaminant mass that amounts to 5% or less. Although the resuspension losses and residuals are small relative to the total mass

---

[11]It should be noted that the resuspended solids fraction measured downstream likely underestimates the total contaminants in the water column because some dissolved releases are likely to occur from solids that remain within the dredge footprint, that is, freshly exposed and redeposited sediments.

dredged, the availability of the contaminants to organisms may be higher than prior to dredging because of exposure to contaminants in the water column (resuspension) or at the sediment surface (residual). The residuals may be especially problematic in that the concentration in the residual is similar to the average concentration in the dredged sediment (Reible et al. 2003; Patmont 2006) and directly accessible to organisms that live at or interact with the sediment-water interface. Because of the presence of the residuals, surface concentrations may not change or may even increase when compared with pre-dredging conditions.

Patmont and Palermo (2007) used a similar (though not identical) dataset to investigate the influence of site-specific factors on residual contamination after dredging. In this analysis, they concluded that

> Similar generated residual percentages were observed for both mechanical and hydraulic dredges. The available data suggest that multiple sources contribute to generated residuals, including resuspension, sloughing, and other factors. However, on a mass basis, sediment resuspension from the dredgehead appears to explain only a portion of the observed generated residuals, suggesting that other sources such as cut slope failure/sloughing are likely quantitatively more important. The available mass balance data also indicate that the presence of hardpan/bedrock, debris, and relatively low dry density sediment results in higher generated residuals.

Figure 4-7, from Patmont and Palermo, shows the influence of debris and/or hardpan and sediment bulk density on the estimated residual at 11 dredging project sites. (Figure 4-7 has one less case study than Figure 4-6.) Higher amounts of debris, the presence of hardpan, and low sediment bulk density all contribute to higher generated residuals.

An examination of dredging at the various sites included in Appendix C indicated that resuspension, release, and residuals can all be influenced by site conditions such as those discussed by Patmont and Palermo (2007) and by the manner in which dredging is implemented. The next section discusses the role that each of those may play in limiting dredging effectiveness on the basis of experience at specific sites where dredging was used; taken together, the experience illustrates site-specific conditions or activities that contribute to or limit dredging effectiveness.

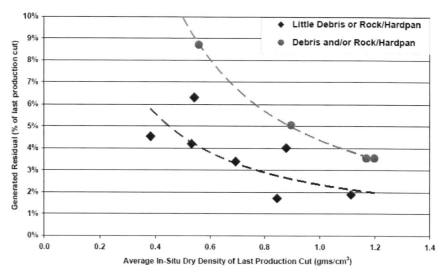

FIGURE 4-7 Estimated generated residuals for 11 projects with data broken down on whether debris, rock, or hardpan was prevalent at the site. Source: Patmont and Palermo 2007. Reprinted with permission from the authors; copyright 2007, Battelle Press.

The objective is to identify conditions under which dredging might be effectively implemented and conditions that could discourage the use of dredging because of its inability to meet desired cleanup levels or remedial action objectives.

Resuspension and Release of Dredged Contaminants During Removal

Resuspension and release of contaminants during dredging are among the most important adverse effects of dredging. As shown in Figure 4-6, up to 10% of the mass of sediment dredged can be resuspended during dredging. Early illustrations of the potential for dredging to give rise to at least short-term increases in adverse effects on organisms can be found in the previously discussed Black River, OH, and Marathon Battery, NY, sites. Those increases were probably due to exposure to more highly contaminated sediment or resuspension of sediment and contaminants during dredging. In the next two sections, the effect of sediment resuspension during dredging and the duration of those effects will be illustrated by experiences at other sites.

Effects of Resuspension and Release

Grasse River, New York

Monitoring of dissolved PCB concentrations in the water column during the 1995 non-time-critical removal in the Grasse River, Massena, NY indicated water concentrations above the 2 µg/L PCB action level at several time points and as high as 13.3 µg/L PCBs (BBL 1995).[12] These concentrations were detected along the perimeter of the project, beyond three lines of silt curtains that were used in an effort to contain suspended sediment. The interior curtain was extended to the bottom, and, as stated in the documentation report (BBL 1995), "the lowering of both the boulder and the inner, secondary curtains to the River bottom greatly enhanced TSS [total suspended solids] containment." This site also used caged fish before, during, and after dredging to monitor bioaccumulation of PCBs (BBL 1995). The study concluded that the increases in caged fish exposed for six weeks during dredging had increases in PCB concentrations 20 to 50 times higher than those observed in the pre-dredging time frame and increases of that magnitude suggest that uptake of PCBs was affected by the release of PCBs to the water column during dredging. However, the report also states that some of the increases in caged fish may be attributable to higher water temperatures during the dredging exposures.[13]

---

[12]In contrast, site-wide sampling the year prior to the dredging generally showed no quantifiable PCB concentrations throughout the river (that is, concentrations were less than 0.5 or 0.7 µg/L depending on the Arochlor) (BBL 1995).

[13]The analysis is further complicated because the increases in fish upstream of the removal were proportionally greater than those downstream. For the 6-week exposures during dredging, lipid-normalized PCB concentrations at the two upstream cages were on average 59 times higher than pre-dredging exposures, while downstream cage locations were on average 35 times higher. However, on an absolute basis, the increases were less pronounced upstream (average increase of 658 ppm PCBs on a lipid basis) relative to the increases observed at the downstream locations (average increase of 2,394 ppm PCBs on a lipid basis). It is suspected that the greater proportional increases in the upstream cages (150 ft upstream of the removal area) resulted because the pre-dredging concentrations in the upstream cages were lower and, thus, increases in PCBs at both locations will result in a greater proportional increase at the upstream cage locations and be-

A second dredging project took place in the Grasse River in 2005. This demonstration project was intended to remove approximately 64,000 cy of PCB-contaminated sediment from 3 work zones, but ultimately removed about 24,600 cy of sediments from approximately 40% of the targeted area. Water column sampling during this demonstration project showed about one-fourth of the 100-odd measurements taken over the course of the project exceeded the 2 µg/L PCB action level in the most downstream sampling site adjacent to the silt curtain (EPA, unpublished data, 4/18/2006). All measured water column concentrations throughout the river prior to the demonstration dredging project were less than 0.065 µg/L total PCBs.

As shown in Figure 4-8, increases in water-column PCB concentration were noted downstream of the dredging operation. TSS increased in concert with PCBs at the sampling locations adjacent to the site, but downstream PCB concentrations remain high when TSS decreased back to upstream values. The presence of dissolved PCBs that would not settle may account for the different behavior downstream. Because dredging is a technology designed to remove solids, dissolved contaminants are contained much less effectively. Similar behavior would be expected for remedial dredging of sediments containing fluid contaminants, such as nonaqueous-phase liquids. In addition, operational controls on resuspension of sediment particles are expected to have less effect on dissolved-phase or fluid-phase contaminants. Silt curtains, for example, are designed to provide additional residence time to encourage particle settling and would have less influence on non-settleable contaminants.

The increase in suspended solids and their flux down river was not associated with high-flow events and therefore was likely due to dredging-related resuspension processes. A horizontal auger dredge was used at this site for most of the demonstration dredging. A horizontal auger of the size used is limited in its ability to dredge effectively in the presence of stone or debris 4 in. or larger. A separate debris-removal operation was implemented to eliminate larger stones and debris. As is typical, an open bucket was used to allow sediment captured with the debris and stone to redeposit on the bottom. Such operations can increase the entry

cause low flows during the removal created conditions favorable for the upstream movement of water (J. Quadrini, written commun., March 23, 2007).

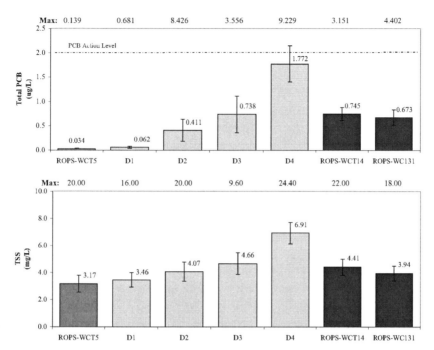

FIGURE 4-8 Average water-column PCBs and TSS during debris and sediment removal activities (conducted simultaneously) in the Grasse River during 2005 demonstration dredging program. ROPS-WCT5 is a sampling site upstream of dredging area; sampling locations D1 through D4 are adjacent to the dredging site in an upstream (D1) to downstream (D4) direction; ROPS-WCT14 and ROPS-WC131 sites are about 0.5 and 1.0 miles downstream of the dredging activity, respectively. Error bars represent ±2 standard errors of the mean. Source: Connolly et al. 2006. Reprinted with permission from the authors; copyright 2006, Quantitative Environmental Analysis.

of suspended solids into the water column and thus increase the contaminant burden in the water column. Similar problems occur in the use of clamshell-bucket dredging of sites laden with debris. The occasional inability to close the bucket completely because of debris interference can increase resuspended solids and thus resuspension of contaminants.

A more detailed depiction of the PCB concentrations in water seen approximately 0.5 miles downstream of the dredge site is presented in Figure 4-9. The pre-dredging baseline PCB concentrations are low; during dredging activities, these concentrations generally increase and occasionally exceed the 2 μg/L PCB action level; after dredging, concentrations decrease back to baseline concentrations (see Figure 4-9). It is estimated that during dredging activities about 3% of the PCBs removed from the river bottom were released downriver, largely as PCBs that had desorbed from resuspended sediments (Connolly et al. 2007).

The overall effect of resuspension and release during the dredging operation can be seen more clearly by examining PCB concentrations in fish at the Grasse River site. Figure 4-10 shows PCB concentrations in spottail shiner measured every fall from 1998 to 2006.

Spottail shiners are useful indicators because they forage only over a limited area (Becker 1983) and, being small, respond quickly to increases in PCB concentrations (Connolly et al. 2006). During 2005, the fish sampling coincided with dredging activities and PCB concentrations in the spottail shiner increased dramatically. The following year, PCB concentrations in shiners decreased to levels seen prior to dredging. There was a statistically significant increase in the downstream locations, but there is insufficient information to evaluate trends associated with dredging, because only a single post-dredging monitoring period was

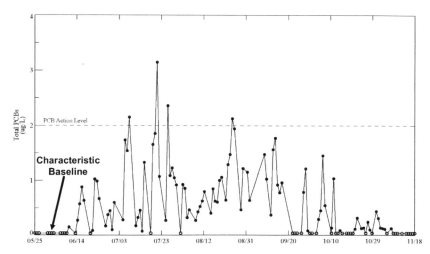

FIGURE 4-9 PCB concentrations in water samples collected approximately 0.5 miles downstream of the dredging operations in the Grasse River (NY). Dredging began approximately June 8, 2005 and ended October 21, 2005. Source: Connolly et al. 2006. Reprinted with permission from the authors; copyright 2006, Quantitative Environmental Analysis.

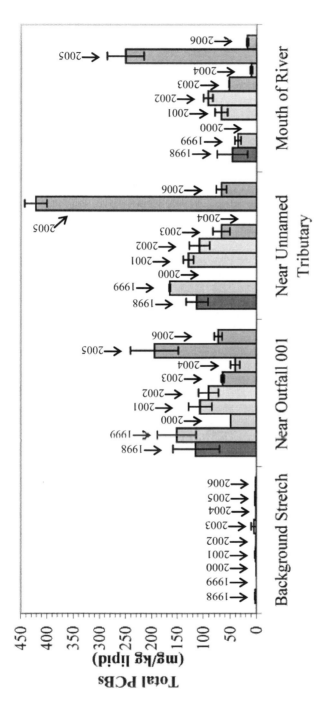

FIGURE 4-10 Lipid-normalized PCB concentration in resident spottail shiners in Grasse River, Massena, NY. Locations are from upstream (Background and Near Outfall 001) to downstream (Near Unnamed Tributary [about 2 miles downstream] and Mouth of River [about 6 miles downstream]). Error bars represent ±2 standard errors of the mean. Dredging at the site took place June to October, 2005; the 2005 fish tissue sampling was in September, 2005. Source: L. McShea, Alcoa Inc., unpublished material, February 27, 2007.

available for PCB body-burden analysis (see Box 4-1). Additional data collection and detailed analyses of the Grasse River data are ongoing by EPA and Alcoa.

Duwamish Diagonal, Washington

In the Duwamish/Diagonal CSO, WA, early-action sediment dredging project, high PCB concentrations (above the pre-dredging surface concentrations) were found in sediments in and outside the dredge prism during a pre- and post-dredging sampling program (EcoChem Inc. 2005). During dredging, several complaints were logged about poor dredging practices that may have contributed to resuspension and release of contaminants. The dredged areas were capped, and to address the unexpected contamination created outside the dredge prism, 6-9 in. of sand was added to areas adjacent to the dredging area. Water quality monitoring during dredging indicated that turbidity standards were exceeded on several occasions (particularly in the first 2 weeks). Total PCBs and dissolved mercury (measured only during the first 8 days of dredging) were below water quality standards even at the highest turbidity values (EcoChem Inc. 2005).

As part of the overall RI/FS of the Lower Duwamish River Superfund site and other studies, fish tissue samples were collected both before and after the Duwamish/Diagonal dredging project. These data suggest that fish-tissue PCB concentrations were greater after dredging activities (Figure 4-11). While there is only one data point for the targeted species collected several years prior to dredging, other fish data collected in the project area in the years prior to the early action (that is, between 1992 and 2000) indicate that tissue levels remained steady during this period (J. Stern, King County Department of Natural Resources and Parks, personal commun., April 20, 2007). The exposure dynamics of resident fish are complex, so monitoring data alone are unable to directly implicate dredging as the cause of the apparent "spike" in tissue PCB concentrations. However, the timing of the increased fish concentrations, the rapid decrease after dredging, and corroborating fish bioaccumulation modeling (Patmont 2006; Stern and Patmont 2006; Stern et al. 2007) are suggestive of a dredging-related release.

BOX 4-1 Statistical Analysis of Fish PCB Body Burdens,
Grasse River, New York

Data

The data analyzed for this report (EPA, unpublished data, 4/18/2005) included percent lipid and total PCB concentrations in fish tissue for smallmouth bass (fillets), brown bullhead (fillets), and spottail shiner (whole body) sampled from the years 1993-2004 (pre-dredging), 2005 (during dredging) and 2006 (post-dredging). Brown bullheads and smallmouth bass were sampled from 4 areas: Background (upstream of dredging) and the Upper, Middle, and Lower stretch of the river (increasing distance downstream from dredging). Spottail shiners were sampled from the Background, Near Outfall 001, Near Unnamed Tributary, River Mouth areas (see Figure 4-10 legend for spottail shiner location details).

Methods

Temporal trends in fish tissue PCB concentrations and region specific effects were established based on linear regression models using monitoring year, percent lipid content, and sampling region as independent variables. PCB concentrations were centered upon their sampling region mean. The Box-Cox transformation (Box and Cox 1964) was parameterized in the regression model likelihoods to allow the data to optimally choose possible transformations. Analyses were stratified by fish species. Nonlinear trends in time were considered (Stow et al. 1995), the results of which led to interpretations that were qualitatively similar.

Different detection limits were reported for these data. Regression model inference was based on maximum likelihood treating the below detection limit data as left censored (Helsel 2005). (Data are left censored if their numeric value only indicates they are less than some given threshold such as the detection limit.) This method has been suggested as an alternative to substitution based techniques such as replacing non-detects with half their detection limit.

Results

Fish tissue PCB concentrations were transformed using natural logarithm for all analyses. Lipid adjusted fish tissue PCBs sampled in the Background region were significantly lower than those sampled in other regions, smallmouth bass ($p<0.01$), brown bullhead ($p<0.01$), spottail shiner ($p<0.01$). Region specific significant decreasing temporal trends based on pre-dredging data (1993-2004) were established for all species.

(Continued on next page)

BOX 4-1 Continued

We explored whether PCB concentrations during dredging (2005) were significantly greater than that expected from the temporal trend established on the pre-dredging data (1993-2004). The post-dredging (2006) data were also compared to the established pre-dredging temporal trend.

For the brown bullhead, the during dredging lipid adjusted average PCB concentrations were larger than the upper 95% limit of the time trend prediction interval for all regions. Results for the Background stretch region is interpreted with caution due to small sample size. For the smallmouth bass, the during dredging lipid adjusted average PCB concentrations were larger than the upper 95% limit of the time trend prediction interval for all regions, except for the Background stretch. For the spottail shiner, the during dredging lipid adjusted average PCB concentrations were larger than the upper 95% limit of the time trend prediction interval for "Mouth of river" and "Near unnamed tributary" areas. Concentrations during dredging in the background and "Near outfall" stretches were within these time trend based prediction limits. Due to small sample sizes results for spottail shiners are interpreted with caution.

The post-dredging (2006) lipid adjusted average PCB concentrations decreased significantly compared to those measured during dredging (2005) for all fish species and all sites, except the background regions for smallmouth bass and brown bullhead. The lipid adjusted fish tissue PCB concentrations post dredging (2006) were within the range (95% prediction intervals) predicted by the temporal trend established on the pre-dredging data (1993-2004) for all species and regions except for smallmouth bass in the lower and middle regions which remained above the 95% prediction intervals.

Remarks

Additional post-dredging sampling points will be needed to evaluate long-term dredging effectiveness. Longitudinal monitoring (encompassing pre- and post-dredging time frames) can be used to statistically compare trends in contaminant concentrations before and after dredging and better associate changes with dredging.

Fox River, Wisconsin

A further illustration of the influence of sediment resuspension on dredging effectiveness can be found in various demonstration projects conducted at the Fox River, WI. At Deposit N, silt curtains were used in 1998 to contain any resuspended sediment, and downstream turbidity

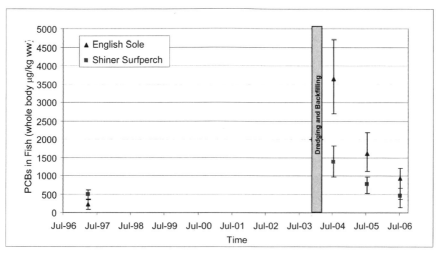

FIGURE 4-11 PCB concentrations in fish collected at the Duwamish Diagonal site pre- and post-dredging. Two trophic levels of fish (English sole and shiner surfperch) of consistent size range were collected within 1 km of the site at the time points indicated on the figure. Error bars represent the minimum and maximum values. Dredging was conducted at the site between November 14, 2003 and January 20, 2004. Capping was completed by February 29, 2004. Sources: Adapted from Patmont 2006; Stern and Patmont 2006; Stern et al. 2007. Fish data from King County 1999; King County, unpublished data, 2007; Windward 2005, 2006.

was found to be no greater than that at upstream sampling locations. Upstream and downstream turbidity levels were also very similar in 1999, when silt curtains were not used (Foth and Van Dyke, 2000). Nevertheless, during 1998 dredging, water-column PCB concentrations averaged 11 ng/L downstream of dredging and 3.2 ng/L upstream. Prior to dredging, the average PCB water column concentrations were similar upstream and downstream of the dredging area. Similar increases (24 ng/L downstream vs 14 ng/L upstream) were observed in 1999. Overall, about 4% of the PCB mass removed from the deposit was released to the water column by dredging (Malcom-Pirnie and TAMS, 2004). The demonstration conducted in Sediment Management Units 56 and 57 (SMU 56/57) in 1999 and 2000 targeted a deposit of about 80,000 cy near the outfall of a recycling mill in Operable Unit (OU) 4, a relatively low-energy estuarine reach of the river. During dredging in 1999, in which

31,000 cy were removed, about 2.2% of the PCBs dredged were estimated to have been resuspended and transported downstream (Steuer 2000).

Direct estimates of contaminant release can rarely be used to guide day to day dredging operations because water column samples may take several days to analyze. Suspended-solids measurements are sometimes used to provide real-time feedback for the dredging operation. As indicated in Box 4-2, however, the correlation between suspended solids and even strongly sorbing contaminants, such as PCBs, may not be adequate to guide operations appropriately.

BOX 4-2 Correlations between Suspended Solids and
Contaminant Concentrations

Although resuspension of sediment is largely viewed as the source of contaminant losses, any contaminant that partitions rapidly from the sediment to the water column will quickly cease to be related to resuspended-sediment concentrations. As shown by the Grasse River 2005 data (Figure 4-8), the water-column concentrations did not generally correlate with suspended-solids concentrations. Turbidity is typically used as a rapid surrogate measure of suspended solids and can be useful to indicate suspended solids if a site-specific correlation between the two quantities can be found. As suggested in the text, however, turbidity and contaminant concentration might not correlate. In an effort to understand the nature of the PCB releases from the Grasse River site, samples were taken from adjacent areas in and outside the silt-containment device surrounding the dredge area. Analysis indicated that although TSS concentrations were about 2 times greater inside the curtain, the dissolved-PCB concentrations were the same (Connolly et al. 2006). At the downstream sampling site, about 0.5 mile from the dredging area, about 75% of the PCBs was dissolved (this is operationally defined as passing a 0.45-μm filter).

At the Fox River SMU 56/57 sites, dredging-related releases resulted in an increase in downstream dissolved-PCB concentrations of about 59%. However, little or no difference in turbidity or TSS concentrations between upstream and downstream locations was detected during dredging, and turbidity and TSS did not correlate with water-column PCB concentrations (Steuer 2000). Those results indicate that turbidity and TSS were of little value as indicators of water-column PCB release. Dissolved and fluid contaminants, such as nonaqueous-phase liquids, will not be well characterized by TSS or turbidity monitoring. A good correlation between suspended solids and contaminant resuspension would be expected if the contaminant remained strongly associated with the solid phase.

Duration of Effects from Resuspension and Release

Increases in contaminant concentrations in the water column and fish during or immediately after dredging may be short-lived. As shown in the Black River, cancer prevalence at the site increased following remediation, but then declined dramatically shortly afterward (Figure 4-1) (Baumann 2000). At Marathon Battery, cadmium concentrations in benthic invertebrates increased following dredging, before declining again (Figure 4-5). In addition, PCB concentrations in the fish in the Grasse River declined substantially a year after concentrations spiked during dredging in 2005 (Figure 4-10).

There is also evidence that contaminant releases from dredging can be reduced. At the GM Massena project in the St. Lawrence River, a sheet-pile wall was erected around the dredging zone because silt curtains were unable to withstand the river currents. The interlocking steel sheet piling enclosed the area to reduce offsite migration of sediment (EPA 2005a). Sheet piling was possible, however, because isolation was required for only a small portion of the river bottom. During dredging, turbidity and PCB and PAH concentrations were measured downstream of the sheet piling to monitor for potential releases into the river. PAH results were consistently below detection limits, and sampling ceased after 19 days. PCBs were monitored over about 3 months. All samples at the monitoring stations were well below the action level of 2 µg/L; the maximum was 0.32 µg/L, and most of the samples were below the detection limit (BBL 1996).[14]

The examples cited indicate that resuspension and contaminant release during dredging can limit at least short-term dredging effectiveness. Large dredging projects that may continue for years or decades are more likely to exhibit more serious problems associated with resuspension and contaminant release than the projects of shorter duration that were examined. For example, dredging could continue for over 25 years

---

[14]The sheet piling may have increased potential concerns about residual contamination while decreasing resuspension losses. In the sheet piled area with the most heavily contaminated sediment, PCB concentrations in water were greater than 20 µg/L a week after dredging stopped. A week later, the concentrations had declined to below 2 µg/L, and the sheet piling was removed. When the sheet pile had been removed, water concentrations dropped to near the detection limit.

at New Bedford Harbor (Dickerson and Brown 2006) and thus any effects of resuspension would be at least of similar duration. There is no experience with such large projects, although some have been initiated (for example, New Bedford Harbor, MA, and Fox River, WI) or planned (for example, Hudson River, NY). The rate of recovery and time to achieve remedial goals after long-term exposure to remedial dredging are not known. Careful design of a monitoring program is needed to separate short-term from longer-term performance of a remedy.

Generation and Exposure of Residual Contamination

Potentially the most serious limitation to dredging effectiveness is residual contamination that is left after dredging. As described in Chapter 2, there are two general types of residuals: generated residuals, contaminated sediment that is resuspended during dredging and later redeposited; and undisturbed residuals, contaminated sediments found at the post-dredge sediment surface that have been uncovered but not fully removed as a result of the dredging operation (Bridges et al. in press). A portion of the generated residual may be unconsolidated and potentially more susceptible to transport. As such, this portion may not be accounted for by confirmation sampling conducted to define post-dredging residuals, depending upon the timing of that sampling. The presence of such residuals directly limits the ability to meet cleanup levels and may also reduce or eliminate opportunities to achieve long-term remedial action objectives. Findings from several of the studied sites on the extent of residual contamination after dredging are provided below.

Lavaca Bay, Texas

A dredging demonstration project was conducted in August 1998 to evaluate the use of a full-size hydraulic dredge to remove mercury-contaminated sediment near the outfall of a chloro-alkali manufacturing facility on the northwest shore of Lavaca Bay, TX.[15] This hot-spot area

---

[15]This description refers to "Phase 1" dredging of the treatability study (Alcoa 2000). The second phase of the study targeted a smaller, less-contaminated shallow water area.

had the highest sediment mercury concentrations in Lavaca Bay, which has widespread mercury contamination. Six acres of very soft plastic clay sediment was dredged in 20 days; 2,300 lbs of mercury was removed with 60,000 to 80,000 cy of sediment, and placed in a confined disposal facility (Alcoa 2000). Extensive data on sediment contaminant concentrations (including post-dredging residual sediment in the dredge area), water quality (including TSS, turbidity and mercury concentrations), and mercury accumulation in caged oysters were collected at the site. Low TSS and insignificant mercury were mobilized beyond the curtained-off zone surrounding the dredge unit. Elevated mercury concentrations in oysters were within the range of those observed in oysters native to Lavaca Bay (Alcoa 2000).

The demonstration project is informative because of the efforts to control sediment residuals. Multiple passes of the dredging operation were conducted with sampling between passes to define the residual present after each pass. There was a notable increase in residual concentration between passes 2 and 3, as shown in Figure 4-12, apparently reflecting exposure of more highly contaminated sediment. Overall, the pass-to-pass concentration changes were not statistically significant, although analyses stratified by subarea did reveal some pass-to-pass concentration changes that were statistically significant (see Box 4-3 and Figure 4-12).

Grasse River, Massena, New York

At the previously discussed 1995 non-time-critical removal action in the Grasse River, Massena, NY, the average PCB concentrations in surficial sediments (upper 8 in.) were reduced by only 53% despite removal of as much as 98% of the PCB mass from the sediment column (Thibodeaux and Duckworth 1999). The site contains numerous rocks and boulders that contributed to residual contamination and contained high concentrations of PCBs near the bottom of the sediment column that could not feasibly be dredged, because of underlying bedrock and glacial till (hardpan) (see Box 4-4).

BOX 4-3 Statistical Analysis of Mercury Concentrations in
Surficial Sediment, Lavaca Bay, Texas

Data

Surficial sediment mercury concentrations in Lavaca Bay (Alcoa 2000) before
any dredging and after four sequential dredging passes were analyzed. Sample locations were identified in four subareas: Capa, North Capa, AA, and the Trench Wall.

Methods

Statistical comparisons of surficial mercury concentrations were based on a linear regression model with indicators of dredge pass (before and after dredging passes
1, 2, 3, and 4) as the independent variables. The Box-Cox transformation (Box and
Cox 1964) was parameterized in the regression-model likelihoods to allow possible
transformations to be chosen optimally. Regression-model inference was based on
maximum likelihood. Stratified analyses were performed for different subareas on
the basis of the Kruskal-Wallis rank sum test (Hollander and Wolfe 1973) because the
samples were small.

Results

Figure 4-12 (left) displays mean surficial mercury concentrations prior to
dredging and after each dredging pass and their corresponding 95% confidence intervals on the log base 10 scale. Mean surficial concentrations after each dredging
pass were not significantly different from the means sampled before dredging. Figure
4-12 (right) displays the distribution of surficial mercury concentrations and corresponding sample sizes stratified by subarea and dredge pass. For Capa, mean surficial mercury concentrations before dredging after dredging passes 1 and 2 did not
differ significantly (p = 0.28). For the Trench Wall, mean surficial mercury concentrations increased sequentially after dredging passes 1, 2, and 3, however these changes
were not statistically significant. The mean concentration after dredging pass 4 was
significantly lower than after dredging pass 3 for the Trench Wall (p = 0.04). Before
dredging, mean surficial mercury concentrations differed significantly among the
four subareas (p = 0.06).

Remarks

Exploratory and followup stratified analysis suggested the geographic subareas of Capa, North Capa, AA, and the Trench Wall as a potential source of surficial
mercury variation. However, sample sizes were insufficient to include this spatial
effect in the regression model while considering variation across dredging times.

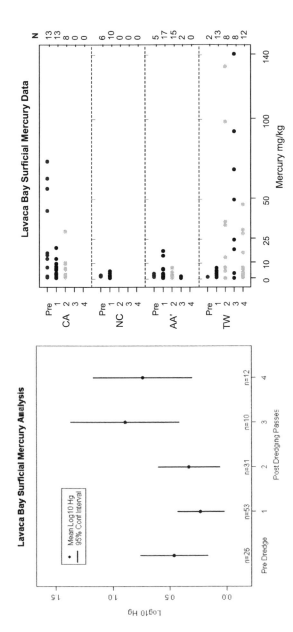

FIGURE 4-12 *Left*, mean and 95% confidence intervals of surficial mercury concentrations (mg/kg) (log base 10 scale) before dredging and after each of four dredging passes. *Right*, concentration distribution and sample sizes (N) of surficial mercury stratified by geographic region—Capa (CA), North Capa (NC), AA, and the Trench Wall (TW)—and dredging pass (Pre, 1, 2, 3, 4; pre = pre-dredging).

BOX 4-4 Description of the Substrate Topography of the
Grasse River During Dredging

"The river bottom, we believe, contains boulders and rock outcrops that
account for these features [seen in the side scan sonar images]. Soft sediment is
intermixed with these features. The river was dredged in early 1900s using
equipment and techniques that may have included blasting, which left a bottom
littered with rocks and boulders, perhaps some outcrops and glacial till. Soft
sediment began to settle on top of this bottom beginning in 1958 when the power
canal ceased contributing flow to the river. All the accumulating soft sediment
contains PCBs because PCBs were discharged from the late 1950s on, and this
contaminated sediment fell in and amongst the rocks and boulders and finally
covered them. So what we encountered is a bottom littered with rock debris intermixed with soft sediment below which is glacial till. The [horizontal auger
dredge] captured some material but the productivity was very low because the
auger couldn't get down in between the rock debris, so I think that was part of
the problem."

Source: Connolly et al. 2006.

Immediately following dredging at the 2005 demonstration project
at the Grasse River, residual surficial sediment PCB concentrations (0-3
in.) averaged 150 mg/kg (dry weight), compared to a pre-dredging average of 4.1 mg/kg (Connolly et al. 2006). The increase occurred despite the
fact that more than 80% of the PCB mass in the dredging footprint was
estimated to have been removed by the dredging operation. Residuals
(generated and undisturbed) were measured at this site and an average
of about 16 in. of contaminated sediments remained after dredging
(range from 3 to 32 in.) (Connolly et al. 2006). Following dredging, the
dredged area was capped with an average of 1.5 ft of a sand and topsoil
mix. At this site, surficial sediment concentrations increased due to
dredging, although there was a large removal of PCB mass from the
river. The analysis illustrates that dredging can achieve substantial mass
removal but may not reduce surficial sediment concentrations. The data
were also analyzed for spatial correlation to identify and account for any
form of spatial variation in surficial PCB concentrations or mass (see Box
4-5).

BOX 4-5 Statistical Analysis of Surficial Sediment PCB Concentrations, Grasse River, New York

Data

PCB concentrations in sediment (mg/kg dry weight) from samples taken at three times—before dredging, after dredging, and after capping—and from two areas—Main Channel and Northern Near Shore—were analyzed (EPA, unpublished data, April 18, 2006). Longitude and latitude spatial coordinates for samples were also available.

Methods

Linear regression models with an indicator of monitoring time (pre-dredging, post-dredging, and post-capping) were used to statistically compare the sediment PCB concentrations before and after dredging and after capping. The spatial sample design (data coordinates) was not sufficiently consistent between times to consider a repeated-measures-based approach. The Box-Cox transformation (Box and Cox 1964) was parameterized in the regression-model likelihoods to allow possible transformations to be chosen optimally. Regression-model inference was based on maximum likelihood. Analyses were stratified by geographic region and considered PCB concentrations. The nonparametric Kruskal-Wallis Rank Sum Test (Hollander and Wolfe 1973) was used to compare surficial PCB concentrations from the Northern Near Shore because the sample sizes were small.

Results

Figure 4-13 displays the region-specific distribution and sample sizes for log10 surficial PCB concentrations. For sediment in the Main Channel, there was a significant increase in average PCB concentrations after dredging (p <0.01) and post-capping concentrations were not significantly different from averaged pre-dredge concentrations. In the Northern Near Shore, the decline in average PCB concentrations between pre-dredging and post-dredging was not statistically significant. After capping the Northern Near Shore area, the average PCB concentration in surficial sediments was significantly lower than before dredging (p <0.01). Regression analyses were on the natural-logarithm-transformed data.

Remarks

The geographic layout of sample locations was not sufficient to allow statistical models to identify and account for any form of spatial variation (either as a regression

(Continued on next page)

BOX 4-5 Continued

trend or residual dependence) in PCB concentration. Residual spatial dependence is
known as potentially biasing tests of significance (Cressie 1991). As described in this
report, the entire main channel area was not able to be dredged. For the Main Channel analysis, pre-dredging, post-dredging, and post-capping data were compiled for
the dredged area only (referred to as "extended work zone 1"). Pre-dredging data
were compiled from three pre-dredging sampling periods for the dredged area
(termed "Phase II," "January 2004," and "Pre-ROPS," generally top 3 in.). The post-
dredging (collected 10/22/2005; generally top 3 in.) and post-capping data (collected
11/28-29/2005; top 2 in.) from that area were also compiled. Northern Near Shore
samples were compiled from pre-dredging (9/9/2004, top 3 in.), post-dredging
(8/19/2005, top 3 in.), and post-capping (11/29/2005, generally top 2 in.) sampling efforts.

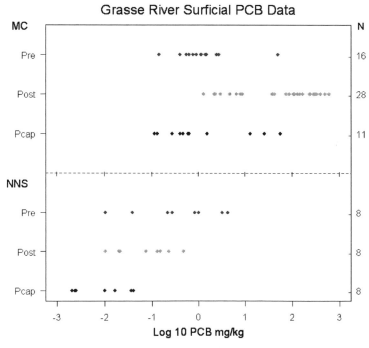

FIGURE 4-13 Distribution and sample sizes for Grasse River surficial sediment
PCB concentrations (mg/kg dry weight) stratified by geographic region—Main
Channel (MC) and Northern Near Shore (NNS)—and whether collected before
dredging (Pre), after dredging (Post), or after capping (Pcap). N = sample number. Further details are provided in Box 4-5.

General Motors Central Foundry, Massena, New York

In 1995, General Motors dredged an area of about 10 acres in the St. Lawrence River near Massena, NY, that was contaminated with PCBs as a result of the release of hydraulic fluids. The action removed over 13,000 cy of sediment and over 99% of the PCB mass in the sediment (EPA 2005a). However, it did not meet the cleanup level of 1 mg/kg in all locations, because there was residual contamination after dredging. As described in the remedial action completion report (BBL 1996), boulders and debris were excavated mechanically, and sediment was removed later with a horizontal auger dredge. The contaminated sediment was underlain with dense glacial till that made it impossible to use over-dredging to increase sediment-removal efficiency. In areas in which initial concentrations exceeded 500 mg/kg, 15-18 dredge passes were required to reduce sediment concentrations to below 500 mg/kg. In one area that initially exceeded 500 mg/kg, eight additional attempts, including multiple dredge passes, were conducted to reduce sediment concentrations. Ultimately, the contractor concluded, with EPA concurrence, that attainment of target cleanup levels in this quadrant was not possible with dredging alone, and capping was instituted (BBL 1996). Without capping, high residual PCB concentrations would have remained at the sediment surface and limited the effectiveness of the remediation.

Manistique Harbor, Michigan

The presence of debris and bedrock limited the effectiveness of the 1995 to 2000 dredging operations to remove PCB contaminated sediment at Manistique Harbor, MI (Nadeau 2006; Weston 2002). The primary remediation goal of the project was the long-term protection of Lake Michigan by removal of the potential PCB source in Manistique Harbor. A secondary goal was reducing risks to people and wildlife that consume fish from the harbor (EPA 1994). As described in Weston (2002), "Initially, the goal of the removal action was to remove sediments within the dredge area with total PCB mass concentrations of more than 10 ppm PCBs …The objectives of the removal action were further clarified and restated …that the "objective of 95% removal of the total PCB mass from within the AOC [Area of Concern] and an average concentration of not more than 10 ppm throughout the sediment column within the AOC shall be verified." Remediation of PCB contaminated sediments began in the fall of 1995 and was originally expected to continue through the fall of 1997 and remove a total of 104,000 cy of sediment (MIDEQ 1996). Ultimately, the project required 6 seasons of dredging from 1995-2000 and removed approximately 190,000 cy of contaminated sediment (EPA 2006a [Manistique River and Harbor Site, May 10, 2006]). The estimated mass of PCBs removed by the end of the project was 82-97% of the initial mass (Weston 2002). Information on pre-remediation surface sediment concentrations varies.[16] Post-dredging average concentrations throughout the river and harbor (including dredged and non-dredged areas) reported for the top 1 ft were 9.0 mg/kg (sampled in 2000) and 7.3 mg/kg (sampled in 2001) (Weston 2002). Nadeau (2006) summarized the pre-

---

[16]EPA describes the results from two sampling events prior to dredging as June 1993: "0-3 in.: Min: 0.15 ppm, Max: 124 ppm, Median: 3.4 ppm" and December 1993: "Min.: Below Detection, Max: 450 ppm" (EPA 2006a [Manistique River and Harbor Site, May 10, 2006]). However, the area being referred to (for example, the whole harbor or just the dredged areas) is not specified. Post-dredging evaluations of site conditions prepared for EPA also do not summarize pre-dredging surface sediment concentrations in the dredged area (Weston 2002; Weston 2005a). A summary by Malcolm Pirnie, Inc. and TAMS Consultants, Inc. (2004) stated that predredging sediment sampling and characterization activities indicated that the average PCB concentration in the top 3 in of sediment was 16.5 ppm. The 1996 Remedial Action Plan for the Manistique Harbor Area of Concern (MIDEQ 1996) states "Sampling conducted in June and December 1993, April 1994 and May, June and July 1995, included most of the navigation channel, along with other harbor and upstream locations...Sampling in the navigation channel showed surface (0" to 3") concentrations of PCBs with a peak value of 120 ppm and an average of 16 ppm." A feasibility study containing a review of experiences at sediment dredging projects (BBL 2000) states "Pre-removal surficial sediment (0-3 in.) PCB concentrations in the Harbor ranged from non-detect to 90 ppm (average of 14 ppm) using data collected during the Engineering Evaluation/Cost Analysis (EE/CA) (BBL, 1994[a]). A consulting report for the responsible party (BBL 1994b) that describes the predredging condition states "The current surficial concentration for the 56 acre area used in the RA is 5.2 ppm not the 8 ppm cited by [EPA]". Several of the above averages are probably derived from the same data source (BBL 1994a), but the areas being considered in that derivation likely differ (for example, the estimated dredged area vs. the 56 acre area of concern).

and post-dredging surface PCB concentrations as increasing slightly over initial surface concentrations (concentrations increased from 5.2 to 7.3 mg/kg in the whole area[17] and 15.1 to 18.8 mg/kg in the dredged area[18]). Since completion of remediation, sediment from upstream has deposited in the harbor area, burying sediments with elevated PCB concentrations. A bathymetric analysis by EPA shows that in a five year period after dredging (between fall 2000 and fall 2005), approximately 83,000 cy of sediment deposited in the harbor from upstream sources, in some places between 10-16 ft deep (EPA 2006a [Manistique River and Harbor Site, May 10, 2006]). Surface sediment samples collected in 2004 (using a ponar dredge) had a mean concentration of 0.88 mg/kg PCBs in the area of interest (Weston 2005a). This example illustrates both the difficulty of eliminating residual sediment concentrations in the presence of debris and bedrock and the inability to achieve long-term risk reduction because of the residual unless other processes, such as sedimentation, intervene to reduce surficial sediment concentrations.

Cumberland Bay, Lake Champlain, Plattsburgh, New York

Debris and a heterogeneous substrate caused dredging problems at the PCB-contaminated Cumberland Bay, NY, site, where logs, wood chips, and large rocks were encountered. Dredging began in July 1999 and ended in December 2000. After the initial dredging of the 34-acre site, divers found many areas where PCB removal was incomplete, apparently because of the presence of debris. As described by the New York State Department of Environmental Conservation (NYSDEC 2001), one area originally thought to have been dredged to a hard bottom was found by divers to be a hard crust that covered 4 ft of sludge containing PCBs at 54 ppm. Hand-held hydraulic dredge lines were used by divers to remove contaminated sediment from areas where dredging with only the large hydraulic dredge was difficult. Another difficulty encountered near a dock area was the bubbling up of gas during dredging that floated sludge to the surface. On the completion of dredging, residual PCB con-

---

[17]5.2 mg/kg: top 3 in., 1993 data, 56 acres; 7.3 mg/kg: top 12 in., 2001 data; 56 acres.

[18]15.1 mg/kg: top 3 in., 1993 data, 15 acre dredged area; 18.8 mg/kg: top 12 in., 2001 data, 15 acre dredged area.

centrations averaged 6.8 mg/kg on the basis of analyses of 51 samples taken from 42 cores. That is lower than pre-remediation PCB concentrations. A dock area had previously averaged 430 mg/kg, with maximum of 13,000 mg/kg, and other dredged areas had averaged 33 mg/kg. However, the reduction in risk cannot be quantified, because no risk-based numeric remediation goals were selected for the site.

Fox River, Wisconsin

Residuals were noted at three dredging projects on the Fox River. At the 1998-1999 removal at Deposit N, sediment rested on a fractured bedrock surface, so it was not possible for a dredge to cut into a clean underlying layer. Sediment PCB concentrations after dredging averaged 14 mg/kg, similar to the average pre-dredging concentration of 16 mg/kg. It is estimated that of the pre-project 142 lb of PCBs measured at Deposit N, 111 lb was removed, and about 31 lb remained in the residual sediment on the completion of the project (Foth and Van Dyke 2000).

At the 1999 Fox River SMU 56/57 demonstration dredging project, steep side slopes, debris, and underlying clay made it difficult to remove contaminated residuals. Final cleanup dredging passes were performed in four subareas before termination of the 1999 dredging, and post-dredging surface concentrations in three of the four areas were less than pre-dredging concentrations, although the 1-mg/kg target was not generally achieved even with cleanup passes. The overall post-dredging average surficial sediment PCB concentration at the end of 1999 dredging was 73 mg/kg, compared with 4.4 mg/kg before dredging (Montgomery Watson 2001). The surficial sediment concentration in the areas not subject to overdredging exhibited an average concentration of 116 mg/kg and a median of 45 mg/kg, very close to the initial 53 mg/kg average concentration in the deposit (Reible et al. 2003). Dredging of the remaining volume was completed by the responsible party as a removal action in 2000 and achieved an average surficial sediment PCB concentration of 2.6 mg/kg; this was followed by backfilling with a minimum of 6 in. of clean sand.[19] The outcome exceeded closure requirements for the re-

---

[19]According to Fort James Corporation et al. (2001): "The vertical extent of the dredging, as determined by the cleanup objectives, resulted in 28 of the subunits being dredged to cleanup objectives, and two of the subunits dredged to develop

moval action, and surficial concentrations after the incorporation of backfill were not measured (Fort James Corporation et al. 2001).

Early reports of the 2005 full-scale dredging of the upper reaches of the Fox River have indicated problems in achieving cleanup targets (Fox et al. 2006). The reports indicate that dredging did not remove all sediment with PCB exceeding 1 mg/kg and that sand backfilling will be necessary to meet the 0.25-mg/kg surface-area-weighted average concentration end point. High-concentration deposits in thin soft sediment layers overlying stiff clay have made residual contamination difficult to remove. A pilot test conducted in a portion of the dredged area indicated strongly diminishing returns for redredging: doubling the volume removed with the goal of removing all soft sediment above native clay was not sufficient to meet the remedial goal of PCB at 1 mg/kg. Dredging results available at the time of the study were from three subunits of OU 1 (Subunit A, C/D2S, and POG1). The preliminary pre-remediation and post-remediation results—PCB mass removal and surficial (upper 4 in.) PCB concentration—from verification sampling are presented in Table 4-1.

More recently, a specialized dredge (a cutter-less head suction dredge) developed by the remedial contractor was used during the 2006 dredging at the site. The dredge was designed for very thin deposits of sediments over a clay or hard till bottom (including generated residuals). Preliminary results for three dredge management units (about 2 acres combined) show that concentrations well under the 1 mg/kg remedial action level were attained even when initial concentrations ranged from 20 mg/kg to above 50 mg/kg (Green et al. 2007). The conditions (thin deposits over clay or hard till bottom) are considered to be among the most difficult for attaining target cleanup levels.

Commencement Bay, Washington

The combination of the ability to overdredge into clean sediment and the presence of sediment that has minimal debris or other obstacles

---

stable sideslopes for the dredge area. All 28 subunits met the cleanup objective of 10 ppm PCBs or less. Eleven of the subunits have PCB concentrations less than 1 ppm."

TABLE 4-1 Summary of Pre-dredging and Post-dredging Verification Sampling Results (2005) from Three Subunits of Operable Unit 1 in the Lower Fox River

Sub-area	Measure	Post-dredging	Pre-dredging	% reduction
A	PCB mass (kg)	205.5	26.6	87
	Avg. surficial PCB conc. (ppm)	13.3	2.8	79
C/D2S	PCB mass (kg)	24.1	1.1	95
	Avg. surficial PCB conc. (ppm)	7.6	1.0	87
POG1	PCB mass (kg)	36.2	1.3	96
	Avg. surficial PCB conc. (ppm)	13.7	1.8	87

Note: PCB concentrations and mass for dredged area only, not entire subarea.
Source: Fox et al. 2006. Reprinted with permission; copyright 2006, Natural Resource Technology, Inc. and CH2M Hill.

to dredging has led to more manageable residual concentrations during the cleanup of several waterways in Commencement Bay, Tacoma, WA. The 1993-1994 Sitcum Waterway cleanup in Commencement Bay was combined with a redevelopment project by the Port of Tacoma designed to create a capacity to handle deep-draft vessels in its facility. As a result of the desire to increase navigable depth, the dredging plan included removal of sediment to a bottom elevation that exceeded the depth of contamination in open-water areas by at least 2 ft. The ability to over-dredge facilitated removal of contamination in those areas and helped to reduce the impact of contaminated residuals on final sediment quality. Immediately after dredging in 1994, sediment quality objectives (SQOs) had not been achieved in all areas. An additional 2 ft of sediment was dredged from one of the areas, and sampling indicated that concentrations in the area were below the SQOs. In the other areas above the SQOs, natural recovery was determined to be sufficient to meet the remedial action objectives; these areas achieved SQOs in 2003. In 2004, EPA approved the Port of Tacoma's request to end further sediment monitoring (EPA 2006a [Commencement Bay–Sitcum Waterway, April 26, 2006]).

The 2004-2006 Head of Hylebos dredging project also indicated the effectiveness of overdredging to reduce residual contamination. The Hylebos Waterway was originally cut into a broad river delta consisting of native sediments composed of clean and fairly compact silts and sands. After the waterway was established, industry developed along the waterway, and this resulted in industrial-chemical discharges into it. No river was feeding the waterway, so it slowly shoaled in with very fine-grained sediment in the form of "soft black muck" over the natural or native sediment. Characterization of subsurface sediment with core samples showed that the contaminants from the industrial discharges were restricted to the fine-grained surface sediment, whereas the immediately underlying native sediment was not contaminated (Dalton, Olmsted & Fuglevand, Inc. 2006). There was a clear visual difference between the contaminated sediment and the underlying native sediment (compact silts and sands). During dredging, each bucket of material was examined visually by onboard inspectors to ensure that all fine-grained sediment had been removed before moving on to the next area. In that manner, overdredging into clean sediment could indicate that residual contamination was minimal. The ability to differentiate clearly between contaminated and uncontaminated sediment, dredging into the uncontaminated sediment, and the relative lack of debris combined to minimize the residuals.

Pre-dredging (dates unspecified) and post-dredging (August 2004-January 2006) surficial-sediment (top 10 cm) total PCB concentrations from the Head of Hylebos and a few samples identified as post-capping (January 2006) samples were available for analysis (unpublished data; Paul Fuglevand; Dalton, Olmsted & Fuglevand, Inc.; August 22, 2006). Linear regression with an indicator of monitoring time (pre-dredging or post-dredging) was used to statistically compare the sediment PCB concentrations before and after dredging. There was a significant decrease in PCB concentrations in surficial sediment after dredging (p <0.01); the pre-dredging mean was 685.9 µg/kg dry weight (n = 135), and the post-dredging mean was 74.7 µg/kg dry weight (n = 400). Only six samples were identified as post-capping samples, and they ranged in concentration from 36.4 to 847.0 µg/kg dry weight. This site remains one of the few where cleanup levels were obtained by dredging alone (except in a few areas). As stated by EPA, "dredging to expose clean native sediment was successfully completed throughout the entire project area, with the ex-

ception of an under-dock cap (completed in 1998) and a shoreline sub-
tidal cap completed early this year at a location of groundwater discharge with elevated arsenic concentrations. The sediment remediation
project successfully achieved the project SQOs with no residual sediments exceeding the SQOs (except at the two noted capping areas which
were not driven by generated residuals from dredging)" (EPA 2006a
[Commencement Bay - Head of Hylebos, May 17, 2006]). The compara-
tively low initial concentrations in the Hylebos Waterway (and many of
the Puget Sound sites) and the smaller magnitude of difference between
contaminated sediment concentrations and cleanup levels decrease the
potential effect of dredging residuals on achieving cleanup levels. However, generated residuals are derived at least partly from the contaminated material being removed so residuals management remains a critical issue at Pacific Northwest sediment sites.

Harbor Island, Duwamish River, Washington

Harbor Island is another site in the Puget Sound area whose stratigraphy is conducive to dredging (a clear separation between contaminated and native, largely uncontaminated sediment). Residual contamination after dredging may be more important than observed at the Head
of Hylebos or Sitcum waterways, because of extensive debris. The Lockheed Shipyard is on the eastern bank of the West Waterway in the lower
Duwamish River in the Harbor Island Superfund site. The site was the
location of a bridge-building company and then a ship-building and
maintenance facility. At the time of remediation, the site comprised a
failing bulkheaded shoreline; almost the entire nearshore area was covered by docks or marine railways, and an open-water area was immediately adjacent to the federal shipping channel. Extensive debris was present in the underpier and open-water area immediately adjacent to the
pier face. Surface debris included consolidated machine turnings and
other metal debris, cable, concrete blocks, and wood. Much of the site
was covered with thousands of deeply embedded creosote piles that
were slated for removal as part of the remedy. The remedy selected for
the site included dredging and capping of the nearshore area (not all
contaminated sediment would be removed) and removal of the contaminated sediment layer in the open-water area including debris that might

limit dredging effectiveness. Habitat restoration was a major component of the remedy (EPA 2006a [Lockheed Shipyard Sediment OU, May 12, 2006]). Additional subsurface debris was encountered once dredging began, including large concrete pier blocks, broken piles, and the original willow cribbing that was used to contain dredged material during the construction of the West Waterway and Harbor Island around 1900. Debris removal had an important effect on the duration, timing, and cost of the project, and the remedy had to be implemented in two phases over two seasons (because of restricted in-water work periods for the protection of endangered species in the Duwamish River). Phase 1 consisted of pier and railway demolition, bulkhead replacement, and initial debris removal (by dredging). Dredging to complete the debris removal and achieve cleanup levels was conducted as phase 2 from November 2003 to March 2004 and from October 2004 to November 2004 (EPA 2006a [Lockheed Shipyard Sediment OU, May 12, 2006]). Because cleanup levels were not achieved after the first phase of dredging, a thin layer of sediment (6-12 in.) was placed over the dredged area to stabilize the residuals until the next season of dredging, when it was removed as part of the dredged inventory.

Summary

The available project data indicates that sediment resuspension and the generation of residuals represent a nearly universal problem in connection with dredging of contaminated sites. Resuspension can be more of a problem in the presence of debris or other site conditions that interfere with normal dredging operations. In addition, readily desorbable contaminants and fluid contaminants, such as nonaqueous-phase liquids, are unlikely to be effectively captured by the dredge or by common operational controls on resuspension. Nor will such contaminants be adequately characterized by measuring the suspended solids, such as TSS or turbidity.

Low sediment bulk density and the presence of debris and hardpan or bedrock all tend to increase resuspension and residuals. Available data indicate that dredging is most likely to be successful when dredges penetrate into clean sediment layers reducing the amount of generated residuals. At sites where structures, debris, hardpan, or bedrock limit dredging effectiveness, the desired cleanup levels, if based on the attainment of specified chemical concentrations, are unlikely to be met by dredging alone. The inability to attain cleanup levels would presumably translate into an inability to meet both short-term and long-term remedial goals and objectives.

Resuspension appears to result in at least short-term negative impacts on water quality and organisms. Residuals may give rise to longer term negative impacts, but at most sites, there has been insufficient monitoring to evaluate the long-term impact of residuals or capping has been used to manage residuals.

MANAGEMENT OF DESIGN AND IMPLEMENTATION TO MAXIMIZE DREDGING EFFECTIVENESS

Although the factors affecting dredging effectiveness outlined in the preceding section are operative at all sites, their influence can be minimized, although not eliminated, by active management of the dredging process, that is, managing design and implementation to maximize effectiveness. Through experience gained at dredging sites, a number of actions have been identified that can help to maximize the effectiveness of dredging in particular situations. However, that experience also suggests that successfully overcoming the limitations of dredging requires both site conditions conducive to dredging and the implementation of some or all of those actions. Sites that exhibit extensive debris, hardpan or bedrock immediately below contamination, or other factors that limit the ability to control residuals or resuspension will continue to be problematic for dredging even if all the actions discussed below are implemented.

Ensure Adequate Site Characterization

Central to the successful implementation of any remedial action is site characterization sufficient to define a conceptual site model. A comprehensive conceptual site model should define the contaminants of concern at a site, the spatial distribution of contamination, the processes that describe the change in contamination over time, the human and ecologic exposure routes, and the significance of exposure and risk. Only when these aspects of the model are developed can a remedial effort be designed to respond to risk appropriately and achieve remedial goals. Adequate site characterization can identify potential sources of contamination and provide the data necessary to design an effective remedial program.

At the Reynolds Metals Superfund site on the St. Lawrence River near Massena, NY, pre-dredging site characterization was not adequate to delineate the distribution of the chemicals of concern at the site. Dredge design was based on the assumption that the PCBs were collocated with the other chemicals of concern, PAHs, and total dibenzofurans. However, post-dredging sampling indicated that this was not the case (EPA 2006c). Following dredging, which included redredging several of the areas, it was determined that PCBs were not collocated with PAHs and that about one-third of the 22 acre dredged area contained PAH concentrations above the cleanup level (EPA 2006c).[20] Future remedial activities at this site are currently being decided.

The Head of the Hylebos Waterway was adequately characterized prior to dredging. Historical surface and core samples were used in conjunction with planned studies to determine the horizontal and vertical distribution of contaminants. As described in the remedial action construction report (Dalton, Olmsted & Fuglevand, Inc. 2006), the historical U.S. Army Corps of Engineers (USACE) post-dredging surveys were used to map the interface between the soft black muck and the native bed sediments and to refine the dredge plan. However, core samples were used to confirm the interfaces, and care was taken not to composite core samples across the muck-native interface. Over 100 cores and over 500 surface samples were used to delineate the area and depth for remediation in this approximately 45 acre site. The coring studies were also used to establish that the recent and native sediments were physically, visually, and chemically different from each other and that chemical ex-

---

[20]Dredging to remove PCBs was more successful although one area required post-dredging capping. As described by EPA (2006c): "Despite extensive dredging of the St. Lawrence River, the cleanup goals of 1 mg/kg PCBs…were not achievable in all areas. As a result, a 0.75-acre, 15 cell area, containing a range of PCB concentrations from 11.1 mg/kg PCBs to 120.457 mg/kg, was capped with the first layer of a three-layer cap to achieve the cleanup goal. The remaining exposed sediments average 0.8 mg/kg PCBs within the remaining 255 cells (21 acres), which is below the cleanup goal."

ceedances of the SQOs were only found in the recent sediments (EPA 2006a [Head of Hylebos Waterway, Commencement Bay, May 17, 2006]). During implementation, the designers viewed the deepest historical dredging as a general guide but relied on observations during dredging to establish successful removal of the impacted sediment.

The design of a dredging plan requires interpolation of depth-of-contamination data from sediment core samples. The upstream portions of the Fox River (OU 1), where dredging is currently being conducted, is a challenging site in that regard because it consists of multiple discrete contaminated sediment deposits arising out of local differences in flow regime and relationship to contaminant sources. The method applied by the remedial design team in OU 1 was to develop deterministic interpolations of sediment PCB concentrations for each deposit and then to connect the interpolations at deposit boundaries (CH2M Hill 2005). When that method is applied, the result is a surface of predicted depth of contamination, which can be expected to be most accurate in the neighborhoods of samples used in the interpolation and most uncertain in unsampled locations and at boundaries of deposits.[21] The spatial density of cores is a key component of adequately characterizing sediment distribution (particularly cores that penetrate through the entire deposit). However, there is no single optimum spacing between core samples because the necessary density depends on the heterogeneity of the site deposit and is site specific.

Accurate characterization is particularly challenging in areas where contaminated sediments are underlain by uneven sub-bottom (for example, furrows, gulleys, or depressions). In these areas, deposition of contaminated sediment over time will often fill in low spots to create a relatively flat sediment-water interface, but with marked differences in the underlying depth of contamination (for example, see description in Box 4-5). At the afore-mentioned Cumberland Bay site, variations in the uncontaminated sub-bottom characteristics and topography proved difficult to characterize prior to dredging. According to the NYSDEC, "site characterization and pre-design studies included bathymetric surveys

---

[21]Alternative interpolation methods exist, such as the geostatistical technique of kriging, providing the probability of contamination at various depths at every location, and interpolating continuously across deposits while allowing for deposit-specific effects (Cressie 1991; Goovaerts 1997; Diggle and Ribeiro 2007).

and sludge probing to define the top of the sludge bed, its thickness, and horizontal extent. In addition to sludge coring, divers confirmed the outer extent of the sludge bed in areas where it was too thin to measure by coring." However, following dredging in 1999, it was determined that in one area "originally believed to have been dredged to a hard bottom since the sampling device encountered refusal" the sediment "consisted of a hard crust underlain by up to four feet of [contaminated] sludge." In another area, PCB-contaminated sludge was found in 1-6 ft deep depressions scattered along the bottom of the lake following dredging. Further dredging targeted both of these contaminated areas (NYSDEC 2001).

In the previously described 2005 pilot study in the Grasse River, it also proved difficult to accurately define the thickness of contaminated sediments using available sampling techniques and protocols. Prior to dredging, multibeam bathymetry, sediment probing, sediment coring, and acoustic sub-bottom profiling were used to characterize the site. Depth to hard bottom was estimated using sediment probing on a 25-ft by 25-ft grid with PCBs expected to be present in sediments above the hard bottom (Connolly et al. 2007). Following dredging, vibracore sampling indicated that in some areas the estimated thickness of contaminated sediments was wrong and significant contaminated sediment remained (see Figure 4-14). These results indicated that at this site "sediment probing is not a reliable indicator of the depth of sediment and manual pushcore sample collection is not a reliable indicator of the full depth of PCBs contaminated material and below the deepest contaminated layer" (EPA 2006a [Grasse River Site, April 18, 2006]). Acoustic subsurface sampling at this site was also not successful for detailing sub-bottom characteristics.[22]

A similar situation existed at the Manistique River and Harbor site. During site characterization, the samples taken before the dredging were

---

[22] According to EPA: "Sub-bottom profiling attempted in conjunction with multibeam bathymetry survey on October 22, 2005. Used dual frequency Odom depth sounder to obtain sounding information along several transects situated parallel to direction of river flow. 200 kHz signal reflects off sediment surface and 24 kHz signal penetrates into sediments and bounces off sediment reflectors. No penetration beyond reflections from sediment surface achieved" (EPA 2006a [Grasse River Site, April 18, 2006]).

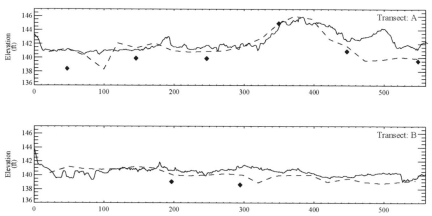

FIGURE 4-14 Upstream to downstream transects of the Grasse River remediation area showing: elevations of the dredging cut line (dotted lines) based on pre-dredging (Spring 2005) depth of contamination estimates; post-dredging (Fall 2005) extent of contamination (diamonds) determined by vibracore sampling; and the post-dredging (Fall 2005) bottom elevation determined by bathymetry (solid lines). Source: Connolly et al. 2007. Reprinted with permission from the authors; copyright 2006, Quantitative Environmental Analysis.

thought to be taken to bedrock, but were not. This is apparently because wood debris under the sediments was thought to be the bedrock harbor bottom (EPA 2006a [Manistique River and Harbor Site, May 10, 2006]). At this a site, a subbottom profiling device was used to estimate sediment thickness. However, the wood pulp and debris in the sediments contained large amounts of gases that rendered the subbottom profiling device useless. As a result, EPA indicated that the dredging depth could not be predetermined in each area (EPA 2006a [Manistique River and Harbor Site, May 10, 2006]).

The influence that incomplete characterization has on the success of dredging points to the importance of accurate site characterization. However, characterization activities are resource intensive and can consume time and funds otherwise available for remedial activities. As a result, decisions on whether to proceed with further characterization should seek to ascertain whether additional characterization will benefit remedial effectiveness and the point at which the additional efforts provide diminishing returns. These considerations will be site-specific. Obviously, areas with complex and heterogenous sub-bottom (such as the

Grasse River) will benefit from greater characterization than those with less heterogeneity. The variety of subsurface characteristics at these sites also indicates that the most useful characterization technologies will be site specific. Sediment sampling coupled with an understanding of the fluvial and geologic nature of an area will shed light on the attributes of the sediment deposit, but not all site conditions can be completely understood prior to beginning work. As a result, verification samples and progress cores taken during dredging are useful for indicating whether operations are succeeding or modifications need to be made (see Chapter 5 for further discussion).

Defining ongoing sources of contaminants to the waterway through site characterization is of critical importance for determining appropriate cleanup responses and for eliminating recontamination of remediated areas. This issue is addressed further in the next section.

Implement Source Control

As pointed out by the National Research Council Committee on Remediation of PCB-Contaminated Sediments, "the identification and adequate control of sources of PCB releases should be an essential early step in the site risk management" (NRC 2001). If contaminant sources are not controlled, dredging cannot be effective in managing risk. In the Hylebos Waterway, the combined efforts of EPA and the Washington State Department of Ecology to achieve source control before dredging contributed to the ability to implement successful remedies. Dredging of the Hylebos Waterway was also at least partially successful because of the efforts to directly address inaccessible areas that could not be dredged. Much of the shoreline of the waterway is modified with over-water structures, such as docks, piers, and wharves. The remedy selected for the head of the waterway included dredging accessible areas, excavation or capping in isolated intertidal and under-pier areas, and natural recovery. Without control of the contaminated sediments under piers and along the shoreline, the project would likely not have been considered successful. Dredging beneath the Arkema dock during 2005 used a long-reach excavator to remove most of the impacted sediment followed by a diver-deployed hydraulic dredge to remove the loose residual material that accumulated during mechanical dredging. The 2005 dredging activities achieved the SQO cleanup objectives beneath both the Ace Tank and Arkema docks. Some areas along the shore were inaccessible to dredging and were either capped or determined to be suitable for monitored natural recovery.

In contrast, both potentially important source areas and exposed sediment contributing to risk were missed, or not appropriately evaluated, in the characterization of the Lauritzen Channel at the United Heckathorn site. Dredging did not achieve remedial action objectives at this site despite achieving cleanup goals immediately after dredging for DDT, the contaminant of concern—apparently because of a failure to address contaminant sources or mass contributing to exposure and risk at the site. Confirmation sampling after dredging seemed to confirm the cleanup goal of 0.59 mg/kg in the Lauritzen Channel of the site, but an investigation a year later, in 1998, found DDT concentrations as high as 30.1 mg/kg. Year 1 biomonitoring showed that pesticide concentrations in the tissues of mussels exposed at the site were higher than those observed before remediation; these values decreased slightly in later years. Anderson et al. (2000) noted remaining toxicity in amphipods after dredging. In 2002, a buried outfall visible during low tide was identified as a persistent source of DDT in the channel; it was plugged by EPA in 2003. Investigations in 2002 and 2003 found sediment concentrations greater than 1,000 mg/kg in the vicinity of a dock on the eastern side of the channel. Remedial objectives were not met in the Lauritzen Channel for a number of reasons, including the decision not to dredge side slopes completely or to remove material from under piers. Some of those areas were intended to be capped with sand, but because of the steepness of slopes or inaccessibility, this was not done (Chemical Waste Management 1997).

A recent analysis of recontamination of completed sediment remedies based on publicly available reports, such as 5 Year Reviews, indicated that 20 areas where dredging or capping remedies had been completed were recontaminated from outside sources, primarily by combined sewer outfalls (CSOs), unremediated upland areas, and adjacent and upstream unremediated areas (Nadeau and Skaggs 2007). The potential for recontamination at a site underscores the importance of identifying and controlling sources before undertaking a sediment remedy.

Monitor Appropriate Indicators of Effectiveness

Adequate site characterization should provide an understanding of the key sources of exposure and risk at a site and of how to intervene effectively to control risk. It should allow the definition of appropriate remedial action objectives and of cleanup levels that will lead to their achievement. It should also therefore identify appropriate indicators of successful implementation and, ultimately, effectiveness. A baseline of appropriate indicators of effectiveness must be established before dredging to make possible a comparison with post-dredging data. Monitoring should continue until effectiveness can be evaluated.

At the Outboard Marine Corporation—Waukegan Harbor site, success of the dredging remedy was monitored by using, among other indicators, fish-tissue data. The site is in Lake County, IL, 50 miles north of Chicago, and consists of industrial, commercial, municipal, and open or vacant lands at the mouth of the Waukegan River and North Ditch drainage basins. It is estimated that 300,000 lb of PCBs (Aroclors 1242 and 1248) were released into the harbor and that sediment concentrations were up to 25,000 ppm (EPA 2000a). The harbor was dredged in 1992 and contaminated sediments were placed in an abandoned boat slip. Only areas exceeding 50 ppm were remediated; as a result, some areas in the harbor are expected to have relatively high residual concentrations, and further remediation is being considered (EPA 2002; EPA 2007b).

Fish-tissue data from Waukegan Harbor were analyzed to evaluate the hypothesis that PCB contamination has decreased (see Box 4-6). To evaluate dredging effectiveness accurately on the basis of pre-remediation and post-remediation fish body burdens, data sufficient to estimate both a pre-dredging time trend and a post-dredging time trend are needed from representative samples of fish collected from exposure areas that are the subject of cleanup levels and remedial action objectives. Effectiveness can then be shown if the post-dredging trend is lower than would be expected from simple extrapolation of the pre-dredging natural-recovery trend. For Waukegan Harbor, sampling was conducted at only two times before dredging. That pre-dredging time trend was inadequate for comparing to the post-dredging time trend.

A pre-dredging sediment toxicity test found relationships between toxicity and sediment-contaminant concentrations, with toxicity ranging

BOX 4-6 Statistical Analysis of PCB Concentrations in Fish,
Waukegan Harbor, Illinois

Data

Analyzed data (T. Hornshaw, Illinois EPA, written commun., August 3, 2006) included percent lipid and total PCB concentrations in carp tissue (fillets) that were caught before dredging (1981 and 1983) and after dredging (1996-2001 and 2005). Dredging activity was conducted during 1991-1992.

Methods

Temporal analysis of fish trends was based on linear-regression models of PCB concentrations in fish samples with monitoring year and percent lipid content as independent variables. The Box-Cox transformation (Box and Cox 1964) was parameterized in the regression-model likelihoods to allow possible transformations to be chosen optimally. Nonlinear trends in time were considered (Stow et al. 1995), and their results led to interpretations that were qualitatively similar. The two-sample Wilcoxon Rank Sum Test (Hollander and Wolfe 1973) was used to compare percent lipid-normalized PCB body burdens before and after dredging.

Results

Figure 4-15 displays the log base 10 lipid-adjusted total PCB in carp fillets for the available monitoring years. There was no significant difference between mean pre-dredging (1981 and 1983) lipid-normalized PCB and mean post-dredging (1996-2001 and 2005) lipid-normalized PCB ($p = 0.34$). Despite the scatter of the data points, there is a statistically significant 12% decline in lipid-adjusted PCB per year for 1996-2005 ($p = 0.03$).

Remarks

The temporal trend shown in Figure 4-15 was established only on the post-dredging data. Comparisons with the pre-dredging monitored fish (total sample size, 7) were insufficient to establish any conclusions on dredging effectiveness. Improved longitudinal monitoring (before and after dredging) could provide data sufficient to establish a time trend that could be used to associate changes with dredging. Monitoring during dredging would provide insight into the effects of sediment release and resuspension and inform statistical models as to when to postulate post-dredging effects.

The comparison of pre- and post-remediation data is further complicated by likely improvements in analytical procedures between 1985 and 2005. Other issues, for example, the failure to segregate fish by age or size, are also likely to affect this analysis.

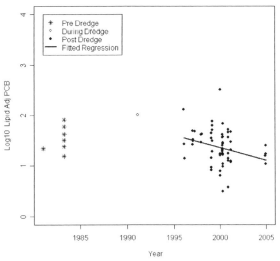

FIGURE 4-15 Lipid-normalized PCB concentrations in carp in Waukegan Harbor, IL. Dredging was completed in 1992. See Box 4-6 for details of the regression analysis.

from 0 to 100% survival for *Ceriodaphnia dubia* and *Daphnia magna* in whole-sediment laboratory assays. *Hyalella azteca* survival was much better in 48-hr exposures with survivals of 73.3 to 100 % (Burton et al. 1989). A post-dredging study by Kemble et al. in 2000 suggested that the PCB concentrations were lower (less than 10 mg/kg) and that sediments were generally not lethal to amphipods, but there were sublethal effects. The Kemble et al. (2000) and EPA (1999) studies suggested that because post-dredging PAH concentrations in sediments exceeded sediment quality guidelines based on probable effect concentrations, they may be contributing to the observed toxicity. The presence of PAH-related toxicity at a site with cleanup levels based on PCBs points to the need to focus on the full range of chemicals that may be causing toxicity at a site.

At the PCB-contaminated Cumberland Bay NPL, NY, site where 34 acres was dredged, quantitative remedial goals were not set. Rather, according to the site's ROD (NYSDEC 1997), "The goals selected for this site are: mitigate the immediate threat to the environment posed by the PCB contaminated sludge bed; rapidly and significantly reduce human health and environmental risks; [and] prevent further environmental

degradation resulting from this known source of PCB contamination."
Without the guidance of risk-based quantitative criteria, it is difficult to
judge whether the remedy achieved the goals that were set. As mentioned previously, PCBs averaging 6.8 mg/kg were still present on completion of dredging.

At the Fox River, a baseline monitoring plan is under development
and is expected to establish the framework for long-term monitoring.
However, the plan is being developed after remedial dredging has begun. A wealth of data is available from the remedial investigation and
from scientific studies of the Lower Fox River and Green Bay that have
been conducted over the decades. Nevertheless, the initiation of the
dredging remedy before the establishment of a baseline—which ideally
would use media, methods, and locations consistent with the long-term
monitoring to follow—will probably complicate the evaluation of remedy effectiveness. In particular, it will be especially difficult to infer the
initial 5-year effects of dredging on contaminant exposures by comparing years 0 and 5, because the apparent baseline will also be affected by
dredging rather than reflect true pre-dredging conditions. Whether
dredging causes an immediate drop in exposures by removing contamination or causes an increase in exposures to contaminated residuals or
water-column releases, the collection of baseline data after the beginning
of dredging will confound the comparison of long-term monitoring data
with the true baseline. That will be true not only of the upstream portion
of the Fox River, where dredging is being implemented, but also of
downstream locations because upstream PCB releases to the water column can affect downstream conditions.

A common problem in monitoring for effectiveness is focusing on
the meeting of cleanup levels, especially if operationally defined, and not
on the long-term remedial action objectives. For example, cleanup for the
Christina River was based on sediment removal. Sediment contaminant
concentrations or biologic responses were not reported immediately following dredging or after backfilling (URS 1999). 5 years after dredging,
data on the status of the benthic community was collected (EPA 2005b);
however, this information is unable to indicate whether the remediation
was effective.[23] Although the cleanup requirements based on dredging to

---

[23]In the second 5-year review for the site (EPA 2005b), EPA states that "the
benthic community is dominated by pollution tolerant species that can be found

a specified elevation may have been met, attainment of short-term and long-term risk-reduction goals has not been demonstrated. Similarly, cleanup requirements were met at the Naval Shipyard in Newport, RI, by removal of sediment to bedrock in many locations (TetraTech 2004). However, dredging to the specified depth (bedrock) is not an appropriate indicator of risk reduction. There was no verification sampling in these locations prior to capping (backfilling), so residual contamination may remain. As a result, contaminants may also be available for transport through the sand cap to surface sediments. The observation of continued toxicity to sea urchins during long-term monitoring suggests that risk-reduction goals were not achieved (TetraTech NUS 2006).

Evidence from other sites shows how meeting of cleanup levels, even if based on post-dredging concentrations, might not achieve desired risk-reduction goals. The 1995 remediation at GM Massena, NY, was designed to ultimately reduce exposure of fish and other wildlife to PCBs in the sediment. As indicated previously, cleanup levels were achieved only after capping of a portion of the site where reduction in residual concentrations below the cleanup level was not achievable solely through dredging. Examination of monitoring designed to evaluate performance relative to long-term remedial action objectives, however, has not shown expected reductions in fish concentrations. As shown in Table 4-2, spottail shiner, a fish with a limited foraging range, showed no obvious increasing or decreasing trends in PCB concentrations at the site even 5 years after the end of remediation (EPA 2005a).

Use Cleanup Levels Appropriately in Determining Remedy Effectiveness

When comparing post-remediation concentration data to cleanup levels, risk managers sometimes treat the cleanup levels as concentrations that should never be exceeded. However, this approach is not nec-

---

in naturally stressed freshwater systems. One of the lines of evidence used to determine that the river needed cleanup was the abundance of pollution tolerant species. Since these areas have been dredged and backfilled with clean sediments, the prevalence of pollution tolerant species would not be attributable to site related contaminant toxicity (there is still zinc in the surface water from other sources)."

essarily appropriate or consistent with the evaluation of human and ecologic exposure conducted in the baseline risk assessments and, more importantly, with the derivation of cleanup levels. EPA guidance (EPA
1989b) recommends use of arithmetic mean concentrations within each
exposure area to quantify exposures to chemicals of concern over time.
While concentrations can vary significantly within an exposure area, the
arithmetic mean is the appropriate statistic based on the assumption that
a receptor integrates its exposure by moving about within the exposure
area. Surface area weighted average concentrations, for example, at the
Fox River, have also been used as appropriate indicators of exposure to
surficial sediments (WI DNR/EPA 2002). The Marathon Battery site illustrates the importance of comparing cleanup levels to the appropriate statistic. At this site, occasional sediment samples had concentrations that
were indistinguishable from the pre-remediation concentration distribution, yet these samples apparently are not reflected in the cadmium body
burden in the benthic community, which has decreased.

Because a sampling program provides an imperfect measure of the
arithmetic mean, EPA recommends use of the 95% upper confidence
limit (UCL) of the mean. Therefore, cleanup levels ideally should be
compared to the 95% UCL for monitoring data representative of the
exposure area of concern that incited establishment of the cleanup level.
EPA used 95% UCLs calculated from surface-sediment samples collected
after completion of dredging to determine whether cleanup levels had
been achieved at the Sitcum Waterway in Puget Sound (EPA 2006a
[Commencement Bay–Sitcum Waterway, April 26, 2006]). For the Sitcum

TABLE 4-2 Spottail Shiner PCB Concentrations After Remediation of the GM
Massena, NY, Site in 1995

Date	Number of Samples	Lipids (%)	Total PCBs-Whole Body Concentration (mg/kg)	Lipid-Normalized PCB (mg/kg-lipid)
10/97	7	5.58	1.20	22
10/98	7	4.24	3.59	79
10/99	7	9.22	2.43	27
10/00	7	11.4	1.5	13
10/01	7	5.00	3.7	75

Source: EPA 2005a.

Waterway site, EPA reported that the "general approach [was to] re-dredge if [concentrations] exceed SQOs, but [EPA] also looked at surrounding data, historical data, and 95th percentile UCL of the mean sediment concentration for a chemical in a subarea to recommend whether additional sampling should be done, re-dredging, natural recovery, and/or nothing." In cases in which monitoring data are biased such that they are not representative of exposure areas of concern, one can perform spatial weighting of the data before calculating 95% UCLs. For example, EPA used an interpolation method called inverse distance weighting to spatially weight floodplain soil data affected by contaminated sediment in the Housatonic River before calculating 95% UCLs for use in the human health risk assessment (Weston 2005b, Attachments 3 and 4).

Consider Using Pilot Tests

As described above, adverse site conditions may significantly limit the ability of dredging to achieve cleanup levels and remedial action objectives. Pilot testing can assist in identifying and characterizing potential limitations to dredging effectiveness, planning and responding to unexpected factors that may arise, and in defining the degree of effectiveness that might be obtainable through dredging. At the Lockheed Shipyard in Puget Sound, delays, additional costs, and limitations of dredging effectiveness were encountered owing to the unexpected quantity of debris. In retrospect, the construction manager for the remediation project stated that a pilot dredging program would have been able to inform him of the extent of the debris and the implications for dredging production rates and rehandling issues (G. Gunderson, TRC Solutions, personal commun., July 7, 2006). Similarly, the dredging remedy at Manistique Harbor, MI, was much more expensive, took twice as long, and required the dredging of nearly twice the expected sediment volume because of the unexpected problems associated with the presence of debris and thin sediment layers over bedrock. A pilot dredging program would have been advantageous in efficiently characterizing the scale and costs of the dredging remedy before full-scale implementation.

Pilot testing has been successfully used at a number of sites in identifying potential limitations to remedial effectiveness and allowing the development of appropriate responses. For example, pilot testing has been used at the Grasse River in Massena, NY, in which the effectiveness of different dredges and different remedial technologies was explored. The previously described demonstration dredging project at Lavaca Bay, TX, was designed to evaluate the ability of dredging to effectively and economically address mercury-contaminated bed sediment at the outfall of a former chloro-alkali manufacturing site (Alcoa 2000). Results of the study indicated that hydraulic dredging could be readily implemented at this site, offsite transport of mercury on tidal flows moving through and around the curtained-off dredging unit were minimal, a large mass of mercury (2,300 lbs) was extracted from the hot spot, and increased mercury concentrations in oysters above the historical observed background in the wider bay did not occur. Residual surface sediment (generally, 0-5 cm) mercury concentrations were reduced in areas with high surface concentrations and lower subsurface concentration, while areas with highly contaminated buried sediments and low pre-dredging surface sediment contaminant concentrations typically showed increased surface sediment concentrations post-dredging (Alcoa 2000). The pilot study was judged to be a successful undertaking in that the data collected were key in the evaluation of the role of dredging in the remedial activities to be undertaken at this site.

Several dredging demonstration projects were conducted during the remedial investigation and feasibility study at the Fox River. The previously-described demonstration projects conducted in Sediment Management Units 56 and 57 (SMU 56/57) in 1999 and 2000 provided valuable information on dredging and dewatering productivity and operations. The 1999 demonstration removed 31,000 cy which was much less than the 80,000 cy objective. During that project, hydraulic dredging equipment was upgraded three times in an effort to increase the solids content of the dredged slurry. Dewatering of solids proved to be a limiting constraint on production rate and required installation of additional filter presses. The average production rate was 294 cy/day, compared with a desired rate of 900 cy/day (Montgomery Watson 2001). Further adjustments in dredging and dewatering equipment, beyond those made by the 1999 project team, were needed in 2000 to remove the remaining 50,000 cy of targeted sediments. The resulting average production rates exceeded a project target of 833 cy/day and reached a peak production rate of 1,599 cy/day on a single day near the completion of the removal action (Fort James Corporation et al. 2001).

Pilot studies also assisted in the success of the Head of Hylebos dredging project. Two pilot studies were used to help to define the scope of such problems as debris, provide large samples for additional testing, and help with selection of equipment and development of the dredging operation plan. One study concentrated on how to remove the "soft black muck drainage water" from the mechanically dredged material at an upland storage facility. About 5 gal of water per cubic yard of sediment, or about 40 lb per 2,200 lb, was associated as free water. The second study focused on rail transport and placement into the offsite landfill and helped to refine the rail-transport program. During the pilot study, the contractor observed the dissociation of physical integrity (strength) of the soft fine-grained sediment after handling, both in the barge and on the bottom of the waterway. The loss of strength was seen as a contributor to formation of a flowable residual layer (fluidized soft mud) on the bottom during dredging that needed to be captured (P. Fuglevand; Dalton, Olmsted & Fuglevand, Inc.; personal commun., July 7, 2006).

In some cases, pilot testing is not required, because of the scale of the dredging project or because of other conditions. For example, the relatively small scale of the U.S. Naval Shipyard site in Newport, RI, and the ability to dredge much of the contaminated material from land combined to make the implementation of dredging favorable. Of 30,000 cy dredged, the vast majority was from the nearshore area (TetraTech NUS 2006). The nearshore materials were removed by a long reach excavator, which was operated on a bay haul road constructed for the project, and were loaded directly into offroad dump trucks (Tetra Tech 2004). Dredging of the remaining elevated offshore material was performed from a barge with a crane equipped with a clamshell bucket and loaded onto an adjacent haul barge. The excavator was much faster and less expensive than the barge-mounted crane, and direct loading into haul trucks minimized handling of material (EPA 2006a [Newport Naval, May 17, 2006]).

In most contaminated sediment megasites, however, the scale and complexity of the sites suggest that pilot studies are appropriate and will assist in reducing limitations of dredging effectiveness. Adaptive management[24] and even pilot testing during implementation may still be necessary to respond to unforeseen problems during implementation. Pilot testing alone will not ensure success; to maximize the project's usefulness, the scope and objectives need to be clearly communicated and monitoring needs to be capable of establishing whether objectives were achieved and the factors that influenced the project's performance.

Implement Best Management Practices

Although it is not a guarantee, the adoption of best management practices (BMPs) will help to ensure appropriate implementation of a remedial project. Best management practices are defined on an activity-specific basis and will depend upon the type of dredging and transport equipment used, the environment in which the dredging takes place, and the process "train" or sequencing of the remedial activities. There are no standardized BMPs for environmental dredging, although "lessons learned" from environmental dredging projects to date suggests that there are BMPs that will likely be applicable to many dredging projects. These BMPs are primarily designed to minimize the loss or transport of contaminated sediment or debris from the dredging footprint and minimizing the generation and runoff of leachate from dredged material to the receiving water during transport or rehandling of dredged sediment. BMPs that may be useful for minimizing loss or off-site transport of sediment and debris include (this list is not considered comprehensive):

- Use of silt curtains to reduce the transport of suspended solids.
- Use of floating and/or absorbent booms to capture floating debris or oil sheens.
- Reduction of the impact speed of the dredge bucket with the bottom and/or reduction of the rate of ascent of a filled bucket; reduction of the swing rate of cutter-head dredge.
- Prevention of overfilling buckets through accurate and controlled placement of bucket.

---

[24]In general, adaptive management is the testing of hypotheses and conclusions and re-evaluation of site assumptions and decisions as new information is gathered. See Chapter 6 for further detail.

- Use of environmental or sealed buckets, where sediment characteristics will allow.
- Protection of the overwater swing path of a filled bucket (by placing an empty barge or apron to catch lost material).
- Eliminating bottom stockpiling of dredged material or sweeping with the dredge bucket/head.

BMPs that may be useful for controlling production or runoff of leachate include:

- Maximization of the "bite" of a dredge bucket (that is, avoiding thin lifts).
- Allowance for draining a sediment-filled bucket before breaking the water's surface.
- Use of filtration cloth, hay bales, curbing, or other physical baffles (similar to stormwater BMPs) to control runoff from barges or rehandling areas.

Additional BMPs that may be used to minimize environmental impacts of dredging include (but are not limited to):

- Scheduling dredging during periods when sensitive species or populations are not present at the site.
- Daily construction oversight and progress surveys.
- Water quality monitoring during dredging activities.

BMPs for control or prevention of resuspension or loss of contaminated material in the waterway were implemented during the remediation of Todd and Lockheed Shipyards. BMPs were specified for overwater demolition, pile removal, dredging, barge dewatering, vessel management, sediment offloading, capping and fill placement, and overwater construction. During demolition, pile removal, and overwater construction, an absorbent boom with 4- to 6-ft silt curtains was deployed to contain floating debris or sheen caused by the removal of creosoted piles. Entrainment of water during dredging was minimized by taking complete "bites" with the dredge bucket whenever possible. Each full bucket was held just at the water's surface to allow water to drain before the bucket was swung to the barge. Dredged sediment was passively dewatered on an onsite flat-deck barge through straw bales and filter fabric before being discharged to the waterway. During offloading, the clamshell bucket was prevented from swinging over open water by placement of a spill-collection platform under its path. Asphalt curbing surrounded the transloading area to prevent sediment, sediment drainage water, and contact stormwater from migrating offsite. Water collected from the transloading area was not allowed to enter the waterway but was collected and treated on site by a process of settling, multimedia filtration, and carbon filtration. Treated water was discharged to the sanitary sewer.

Implementation of BMPs for control of produced water was also important in the success of the Head of Hylebos dredging project. All the water entrained with the dredged sediment was placed in the barge rather than being released back to the water. Overall, the enclosed mechanical buckets placed more water than sediment in the barges. This water can contain an important load of sediments and contaminants. During the 2005 season, the water-management system captured about 4,000 cy of sediment. It is estimated that if that material had been released back to the dredge area, it would have generated a layer of impacted sediment an average of 3-4 in. thick over the dredged area. Capture of the solids contributed to the ability to meet the cleanup goals (Dalton, Olmsted & Fuglevand, Inc. 2006).

Remediation contractors for the Todd Shipyard found benefit in working with dredging companies that were able to mobilize an array of equipment from their inventory to meet project needs and respond to changing site conditions and schedules. The project engineer for Todd Shipyard included the dredging contractor as a consultant during the design stages of the remediation—an important step that ensured a smooth transition from design to implementation (EPA 2006a [Todd Shipyards Sediment OU, May 12, 2006]). Experienced environmental dredgers also have the capabilities to operate within the typical regulatory restrictions and an understanding of the difficulties associated with environmental dredging. The importance of using contractors experienced in environmental dredging was emphasized by the remediation contractors at the Fox River SMU 56/57 dredging project:

Most large dredging contractors in the United States have little or no experience with contaminated sediment projects, working predominantly on navigational dredging projects. Navigational dredging projects typically have no environmental controls, resulting in higher production rates and lower unit costs. Larger-scale projects may also limit the available temporary water treatment and dewatering equipment unless planned well in advance, as well as onshore land space, that are necessary to complete the work in a timely fashion (Montgomery Watson 2001).[25]

Since 1999, the first year of the SMU 56/57 dredging project, the numbers and experience of firms experienced with environmental dredging has increased. However, the need for contractors familiar with the challenges of environmental dredging remains, particularly at large, multi-year megasite projects.

Use Appropriate Contracting Arrangements

The nature of the contracting vehicle used to conduct the work can drive behavior of the contractor and ultimately impact project results. The contracting terms and approaches provide incentives for contractor performance so the contracting approach needs to be aligned with the project's risk reduction goals. For example, the implementation of BMPs can be encouraged or discouraged by the contracting mechanisms used in a remedial project.

EPA described the importance of the contracting mechanism at the Lockheed Shipyard cleanup (EPA 2006a [Lockheed Shipyard Sediment OU, May 11, 2006]):

The primary keys to success for a complex project such as the LSSOU cleanup include (1) a contract where the dredging contractor was not taking the risk and 2) the dredger was guaranteed a daily rate for each activity. Frequently, in the past, environmental dredging contractors have followed the "navigational dredging

---

[25]It is true that navigation projects often have less environmental controls than environmental dredging projects, but navigation dredging projects typically have to comply with water quality certification and disposal requirements of the Clean Water Act or, in the case of ocean disposal, the Marine Protection, Research, and Sanctuaries Act.

model" in terms of contract style and dredging methods. Under a Unit Rate contract the dredger is in a "production dredging mode, which does not work for environmental dredging. Under the Time and Material [contract] the dredger is not penalized for taking the appropriate time to accomplish the task at hand. Because of this the dredger is more likely to comply with BMPs and to take care to minimize loose of material back into the waterway or to cause re-suspension.

Phase 1 dredging at the site resulted in incomplete debris removal, the presence of undredged inventory, and a large amount of residuals at the end of the in-water work window. On review of the approach, the remediation project manager revised the contract mechanism for phase 2 dredging (second season) to use a time-and-materials approach and requested new bids for the work (EPA 2006a [Lockheed Shipyard Sediment OU, May 11, 2006]).

The Todd Shipyard project engineer found that a cost-plus-incentive-fee form of contract worked well by motivating contractors to complete every aspect of the construction in accordance with defined quality objectives at the lowest overall cost. The contract reimbursed the contractor for all direct costs of the work. That removed the financial risk to the contractor and thereby reduced bid costs to cover unknown or unquantifiable risks (EPA 2006a [Harbor Island Todd Shipyards Sediment OU, May 12, 2006]).

In the Head of Hylebos dredging project, a cost-plus-fee contract assisted in the reduction of residual contamination. The engineers and contractors determined that it was in their best interest to minimize the extent of the residual layer by slowing down the production to match "good housekeeping" on the bottom. That contracting mechanism provided the opportunity for the owner to work with the contractor to adjust construction and operations to achieve the project objectives. Contractor oversight was provided by onsite dredge inspectors during each shift. That oversight was used to make decisions regarding whether the target elevations were met (that is, finding the underlying native material) before dredging moved to the next cut. The cab on each dredge was actually expanded to provide a place for the dredge observer to sit side by side with the operator day and night throughout the dredging operation (Dalton, Olmsted & Fuglevand, Inc. 2006).

Overall, there is no single best contract type. These decisions will depend on site conditions and necessary equipment and materials. However, contracting terms and approaches that encourage contractors to focus on achieving cleanup goals and remedial action objectives are best suited for environmental dredging. These arrangements should create incentives for reducing resuspension and residual production, using best management practices, and adjusting the dredging approach to improve chances of meeting cleanup levels and result in cost savings.

Use Operational Controls to Improve Dredging Accuracy

In addition to appropriate design, implementation, and monitoring, technologic approaches can improve dredging efficiency and effectiveness. Some of them have been used successfully at contaminated sediment sites.

A one-of-a-kind specially designed high-technology dredge outfitted with innovative sensors and controls to achieve a 6-in. excavation-cut tolerance was used to extract creosote-contaminated sediment at Bayou Bonfouca, LA. A cutline to the depth associated with a total PAH concentration of 1,300 ppm was established and programmed as an absolute elevation along a 4,000-ft length by using borehole concentration profiles. Maximum contamination depth was 17 ft (average, 10 ft). Logs, concrete, metal objects, and so on were removed with grapple hooks before excavation (EPA 2006a [Bayou Bonfouca Superfund site, May 12, 2006]). The pre-dredging operation and low tidal fluctuations and low stream flow rate (13 ft³/sec) provided a stable dredging platform (spud barge) in the loose, high-organic-matter layer over "harder" unconsolidated inorganic substrate, which all aided in the implementation of the new precision dredging technology. No post-dredging measurements (such as bottom elevations) were taken to evaluate achievement of the bottom cut-line target programmed into the excavator, nor were any sediment analyses performed to verify achievement of a total PAH concentration of less than 1,300 ppm. Targets for volume of dredged material were achieved, however, and that was the primary goal of controlling the excavation depth.

The dredge used at Todd Shipyards was equipped with a positioning system with 20-cm (GPS-controlled) horizontal accuracy. It provided real-time display and tracking of the horizontal and vertical position of the dredge bucket. Digital GPS receivers and a gyrocompass were used to determine real-time horizontal (X and Y) positioning of the derrick barge and the dredge bucket. An electronic tide gauge was used to allow the operator to determine the proper dredge elevation below the water surface accurately. The vertical position of the bucket was combined with the electronic tide-gauge data to determine the bucket elevation (Z). In addition, the dredge-bucket wires were painted in 1-ft increments to provide for a check on the electronically calculated vertical position. The information generated by the positioning system was electronically stored and used to create maps that showed dredging progress, including the degree of overlap between bucket deployments (EPA 2006a [Harbor Island Todd Shipyards Sediment OU, May 12, 2006]). That approach to navigation and positioning of the dredge bucket has been used at a number of sites in Puget Sound.

Precision positioning systems are seeing increasing use. Although the basic technology of dredging has changed little in recent decades, the ability to position the dredge accurately has improved dramatically; in principle, this can improve our ability to remove contaminated sediment accurately and efficiently if sufficient site-characterization data are available. Regardless of the improvements in dredging methods and equipment, the reliability of the equipment and the availability of skilled operators capable of processing and interpreting the data remain challenges.

Consider Backfilling and Capping to Control Residuals

As indicated previously, the factor limiting dredging effectiveness that is the most difficult to manage is high residual contaminant concentrations. Residuals are always detected after dredging and can be relatively high in concentration and typically of the same order as the average concentration in the dredged material (Reible et al. 2003). The magnitude of residuals can be higher in the presence of debris or when site conditions make it infeasible to overdredge into clean material. Even in favorable dredging conditions, however, some degree of residual control is usually necessary to achieve site cleanup standards and to address site remedial action objectives. Generally, control of residuals is achieved by adding backfill or thin-layer capping; this has clear advantages in achieving bulk sediment contaminant concentration targets even if the backfill layer is intermixed with the residual sediments. Although backfill can effectively manage bulk sediment concentrations, the effectiveness of backfill for aiding long-term risk reduction is less well understood. In addition, the advantages of dredging with backfill for residual control relative to a complete capping remedy need to be assessed during remedy evaluation.

At Bayou Bonfouca in Slidell, LA, backfilling was a necessary phase of the overall remedial operation. Excavation to up to 3 m compromised the stability of the unconsolidated material forming the banks along the bayou. Gravel (about 1 ft) over sand (about 1 ft) and additional fill gave support to the sheet-piling-reinforced banks (5,000 ft long) and served to cap the residual PAH contamination of 1,300 ppm (target concentration). Backfilling to the original bottom grade maintained the historical water flow rates and levels needed for recreational and other boating traffic and allowed the bayou to begin natural recovery.

The remediation of the former Ketchikan Pulp Company facility, in Ward Cove in Ketchikan, AK, also involved some backfilling of dredged areas. The facility operated as a dissolving sulfite pulp mill from 1954 until 1997 and discharged untreated sulfite waste liquor (magnesium bisulfite), pulping solids, and bleaching waste (chlorine caustic) into Ward Cove until 1971, with increasing wastewater treatment after that. Mill operations affected sediment by releasing large quantities of organic material (up to 10 ft thick) as byproducts of wood pulping. The organic material altered the physical structure and chemistry of the sediments and thus the type and abundance of benthic organisms. Degradation of the organic-rich pulping byproduct led to anaerobic conditions in the sediment and production of ammonia, sulfide, and 4-methylphenol in quantities that were potentially toxic to benthic organisms (EPA 2006a, Ketchikan Pulp Company; April 26, 2006). Remedial action objectives included reducing toxicity of surface sediments to benthic life and enhancing benthic recolonization. The selected remedy included thin-layer (6-12 in.) placement of clean sand over dredged areas (to less than the full depth of contamination) and undredged areas and monitoring of natural recovery where thin-layer placement was not practicable. Sand backfilling was expected to achieve the remedial objectives by diluting contaminants and organic matter, both of which are associated with benthic toxicity on this site (EPA 2000b).

Remediation was completed in 2001. The long-term monitoring program includes sediment chemical analysis, toxicity testing, and assessment of the benthic macroinvertebrate community. The first round of long-term monitoring in 2004 found that the remedy appeared to have met its objectives in backfilled areas; chemical concentrations were generally below sediment cleanup levels as determined from pre-remediation toxicity testing, survival of benthic test organisms was high, and benthic species diversity and abundance increased relative to the pre-remediation baseline and are similar to reference areas (Exponent 2005). Additional monitoring is planned for 2007, and equally favorable or improved results may lead to a reduction in required monitoring of backfilled areas. In contrast, 2004 monitoring results showed that only one of four natural recovery areas had improvements comparable to those found in the backfilled areas (Exponent 2005; Herrenkohl et al. 2006).

Backfilling has been used at a variety of sites, including the Fox River and several sites in the Puget Sound area. It has been proposed for many sites that have not met or are unlikely to meet cleanup levels after dredging alone. Backfill that is at least about 6-12 in. thick probably forms an effective separation between much of the benthic community that might colonize the top of the backfill layer and the underlying sediment. Thin sand backfill layers (less than 6 in.), however, are of uncertain effectiveness because the low sorptivity of sand means that the benthic community may be exposed to pore water contaminant concentrations similar to that of uncapped sediment. Exposure and risk are often more closely related to pore water concentration than to bulk sediment concentration, and further research or field monitoring is needed to confirm the appropriateness of thin-layer backfilling.

CONCLUSIONS

On the basis of its review of data and experiences at dredging projects, the committee has reached the following conclusions:

- The committee was generally unable to establish whether dredging alone is capable of achieving long-term risk reduction, because o   Monitoring at most sites does not include the full array of measures necessary to evaluate risk.

o   Dredging may have occurred in conjunction with other remedies or natural processes, or insufficient time may have passed to evaluate long-term risk reduction.

o   A systematic compilation of site data necessary to track remedial effectiveness nationally is lacking.

• Dredging remains one of the few options available for the remediation of contaminated sediments and should be considered, with other options, for managing the risks that they pose.

• Dredging is effective for removal of mass, but mass removal alone may not achieve risk-based goals.

• Dredging will likely have at least short-term adverse effects on the water column and biota.

• Dredging effectiveness is limited by resuspension and release of contaminants during dredging and the generation or exposure of residual contamination by dredging. Those limitations are minimized if site conditions are favorable and the remedy is designed and implemented appropriately.

o   Favorable site conditions include

    –   Little or no debris

    –   A visual or physical texture difference or other rapid mechanism for differentiating clean and contaminated sediments.

         Potential for overdredging into clean material.

    –   Low-gradient bottom and side slopes.

    –   Lack of piers and other obstacles.

    –   Site conditions that promote rapid natural attenuation after dredging (for example, through natural deposition).

    –   Absence of non-aqueous-phase liquid or readily desorbable contaminants.

o   Effective design and implementation factors include

    –   Site characterization sufficient to develop a comprehensive conceptual site model and identify adverse site conditions.

    –   Identification and control of sources on a watershed-wide basis.

      –   Use of pilot studies, where appropriate, to identify adverse site conditions and appropriate management responses.

      –   Application of best management practices to control residuals and resuspension (for example, operational controls at the dredge and on produced streams, appropriate equipment selection, and residual control measures).

      –   Contracting and procurement mechanisms to encourage a focus on cleanup levels and remedial action objectives.

      –   Engagement of experienced and innovative environmental-dredging contractors throughout the design and implementation phases of remediation.

- Dredging *alone* is unlikely to be effective in reaching short-term or long-term goals where sites exhibit one or more unfavorable conditions. Where unfavorable conditions exist, increased contaminant resuspension, release, and residual will tend to limit ability to meet cleanup levels and delay the achievement of remedial action objectives unless managed through a combination of remedies or alternative remedies.

RECOMMENDATIONS

- A remedy should be designed to meet long-term risk-reduction goals. The design should be tested by modeling and monitoring the achievement of long-term remedial action objectives.
- Site conditions that influence dredging effectiveness should be recognized during selection, development, and implementation of the remedy. When conditions unfavorable for dredging exist:

      o   Implementation of one or more pilot tests should be considered to identify optimal remedial approaches and assess their effectiveness.

      o   Adverse effects of resuspension, release, and residuals should be forecast and explicitly considered in expectations of risk.

      o   The ability of combination remedies to lessen the adverse effects of residuals should be considered when evaluating the potential effectiveness of dredging.

o   Best management practices should be implemented to minimize effects of adverse dredging conditions.

o   The possibility of adverse dredging conditions that are not anticipated should be recognized and planned for.

• A good baseline assessment coupled with a well-designed long-term monitoring plan should be implemented to permit evaluation of dredging effectiveness.

o   Well-designed pre-dredging and post-dredging monitoring is necessary to establish effectiveness and indicate achievement of remedial action objectives.

o   Monitoring should be conducted to demonstrate achievement of cleanup levels and to confirm that the cleanup levels achieve remedial action objectives.

o   Data from monitoring should be managed and stored in electronic databases accessible for further analysis.

• Further research, including during dredging pilots and full-scale operations, should be conducted to define mechanisms, rates, causes, and effects of dredging residuals and contaminant resuspension.

REFERENCES

AGC (Advanced Geoservices Corporation). 2001. 2000 Annual Report Long-Term Monitoring Program: Marathon Remediation Site. Advanced Geoservices Corporation, Chaddsford, PA. April 2, 2001.

Alcoa Inc. 1995. The Grasse River: Non-Time Critical Removal Action Documentation Report, December 1995. Alcoa [online]. Available: http://www.the grasseriver.com/ppt/Nov20CAP/NTCRA.htm [accessed October 11, 2006].

Alcoa Inc. 2000. Treatability Study for the ALCOA (Point Comfort)/Lavaca Bay Superfund Site, Draft. January 2000. 84pp.

Alcoa Inc. 2005. Remedial Options Pilot Study Work Plan: Grasse River Study Area Massena, New York. Alcoa Inc., Massena, New York. February 11, 2005.

Anderson, B.S., J.W. Hunt, B.M. Phillips, M. Stoelting, J. Becker, R. Fairey, H.M. Puckett, M. Stephenson, R.S. Tjeerdema and M. Martin. 2000. Ecotoxicologic change at a remediated Superfund Site in San Francisco Bay, California, USA. Environ. Toxicol. Chem. 19(4): 879-887.

Baumann, P.C. 2000. Health of Bullhead in an Urban Fishery after Remedial Dredging Final Report - January 31, 2000. Prepared for U.S. Environmental Protection Agency, Great Lakes National Program Office, Chicago IL, by

U.S. Geological Survey Field Research Station, Ohio State University, Columbus, OH [online]. Available: http://www.epa.gov/glnpo/sediment/Bullhead/index.html [accessed Jan. 9, 2006].

BBL (Blasland, Bouck & Lee, Inc.). 1994a. Engineering Evaluation/Cost Analysis. Volume I of II. Manistique Papers Inc. and Edison Sault Electric Company, Manistique, MI. Blasland, Bouck & Lee, Inc., Syracuse, New York. April 1994.

BBL (Blasland, Bouck & Lee, Inc.). 1994b. Response to Issues Raised by USEPA Region 5 in its Removal Action Recommendation and related documents. Prepared for Manistique Papers Inc., and Edison Sault Electric Company, Manistique, MI, by Blasland, Bouck & Lee, Inc., Syracuse, New York. October 1994.

BBL (Blasland, Bouck & Lee, Inc.). 1995. Non-Time-Critical Removal Action: Documentation Report, Vol. 1. Grasse River Study Area, Massena, New York, Aluminum Company of America, Massena, New York. Draft. Blasland, Bouck & Lee, Inc., Syracuse, New York. December 1995.

BBL (Blasland, Bouck & Lee). 1996. St. Lawrence River Sediment Removal Project Remedial Action Completion Report. Prepared for General Motors Powertrain, Massena, New York, by Blasland, Bouck & Lee, Syracuse, New York. June 1996.

BBL (Blasland, Bouck & Lee, Inc). 2000. Feasibility Study Report - Phase I Allied Paper, Inc./Portage Creek/Kalamazoo River Superfund Site, Kalamazoo and Allegan Counties, Michigan, Appendix D. Site Profiles of Sediment Dredging Projects. Draft for State and Federal Review. Blasland, Bouck & Lee, Inc., Syracuse, New York. October 2000 [online]. Available: http://www.deq.state.mi.us/documents/deq-erd-kzoo-FS-apend-d.pdf [accessed Jan. 8, 2006].

Becker, G. 1983. Fishes of Wisconsin. Madison WI: University of Wisconsin Press.

Box, G.E.P., and D.R. Cox. 1964. An analysis of transformations. J. Roy. Stat. Soc. B 26(2):211-252.

Bremle, G., and P. Larsson. 1998. PCB concentration in fish in a river system after remediation of contaminated sediment. Environ. Sci. Technol. 32(22):3491-3495.

Bremle, G., L. Okla, and P. Larsson. 1998a. PCB in water and sediment of a lake after remediation of contaminated sediment. Ambio 27(5):398-403.

Bremle, G., P. Larsson, T. Hammar, A.Helgee, and B. Troedsson. 1998b. PCB in a river system during sediment remediation. Water Air Soil Poll. 107(1-4):237-250.

Bridges, T.S., Ells, S., Hayes, D., Mount, D., Nadeau, S., Palermo, M., Patmont, C., Schroeder, P. in press. The Four Rs of Environmental Dredging: Resuspension, Release, Residual, and Risk. ERDC TR-07-X. U.S. Army Engineer Research and Development Center, Vicksburg, MS.

Burton, G.A., B.L. Stemmer, K.L. Winks, P.E. Ross, and L.C. Burnett. 1989. A multitrophic level evaluation of sediment toxicity in Waukegan and Indiana Harbors. Environ. Toxicol. Chem. 8(11):1057-1066.

Canonie Environmental. 1996. Waukegan Harbor Remedial Action, Waukegan, IL. Construction Completion Report. Canonie Environmental. July 3, 1996.

CH2M Hill. 2001. Five-Year Review Report: Second Five-Year Review Report For Bayou Bonfouca Superfund Site, Slidell, St. Tammany Parish, Louisiana. Prepared by CH2M Hill, for U.S. Environmental Protection Agency, Region 6, Dallas, TX. June 2001 [online]. Available: http://epa.gov/earth1r6/6sf/pdffiles/bayou_bonfouca_5_year.pdf [accessed Jan. 10, 2007].

CH2M Hill. 2005. Lower Fox River Operable Unit 1 Pre-design—Basis of Design. Prepared for WTM I Company, by CH2MHill, Milwaukee, WI. March 2005.

CH2M Hill. 2006. Bayou Bonfouca - Hurricane Impacts Evaluation. Memorandum to Michael Herbert, EPA Region 6, from Renee Ryan and Scot McKinley, CH2M Hill. Project No. 334872.ET.02. January 25, 2006.

Chemical Waste Management. 1997. Completion Report for Marine Remedial Action on the United Heckathorn Superfund Site, Richmond, CA. October 1997.

Connolly, J., V. Chang, and L. McShea. 2006. Grasse River Remedial Options Pilot Study (ROPS) Findings from Dredging Activities. Presentation at the Second Meeting on Sediment Dredging at Superfund Megasites, June 7, 2006, Irvine, CA.

Connolly, J., J. Quadrini, and L. McShea. 2007. Overview of the 2005 Grasse River Remedial Options Pilot Study. Presentation at the 4th International Conference on Remediation of Contaminated Sediments, January 22-25, 2007, Savannah, GA.

Cressie, N.A.C. 1991. Statistics for Spatial Data. New York: Wiley. 900 pp.

Dalton, Olmsted & Fuglevand, Inc. 2006. Remediation Action Construction Report, Part 1: Head of Hylebos Waterway Problem Area Commencement Bay Nearshore/Tideflats Superfund Site Tacoma, Washington, Review Draft, July 21, 2006. Prepared for Head of Hylebos Cleanup Group, Arkema, Inc, General Metals of Tacoma, Inc, by Dalton, Olmsted & Fuglevand, Inc., Kirkland, WA. July 21, 2006.

Dickerson, D., and J. Brown. 2006. Case Study on New Bedford Harbor. Presentation at the First Meeting on Sediment Dredging at Superfund Megasites, March 22, 2006, Washington DC.

Diggle, P.J., and P.J. Ribeiro. 2007. Model-Based Geostatistics. New York: Springer.

EcoChem, Inc. 2005. Duwamish/Diagonal CSO/SD Sediment Remediation Project Closure Report. Prepared for King County Department of Natural Resources and Parks, Elliott Bay/Duwamish Restoration Program Panel, Seattle, WA. Panel Publication 39. July 2005 [online]. Available: http://dnr. metrokc.gov/WTD/duwamish/diagonal.htm [accessed Jan. 3, 2006].

EPA (U.S. Environmental Protection Agency). 1984. Record of Decision: Outboard Marine Corp. EPA ID: ILD000802827 OU 01 Waukegan, IL. EPA/ROD/R05-84/007. Superfund Information Systems, U.S. Environmental Protection Agency [online]. Available: http://www.epa.gov/superfund/ sites/rods/fulltext/r0584007.pdf [accessed Jan. 9, 2007].

EPA (U.S. Environmental Protection Agency). 1987. EPA Superfund Record of Decision: Bayou Bonfouca, EPA ID: LAD980745632, OU 01, Slidell, LA. EPA/ROD/R06-87/019. Superfund Information Systems, U.S. Environmental Protection Agency [online]. Available: http://www.epa.gov/superfund/ sites/rods/fulltext/r0687019.pdf [accessed Aug. 7, 2006].

EPA (U.S. Environmental Protection Agency). 1989a. Record of Decision: Marathon Battery Corp. EPA ID: NYD010959757. OU 02, Cold Springs, NY. EPA/ROD/R02-89/097. Superfund Information Systems, U.S. Environmental Protection Agency [online]. Available: http://www.epa.gov/superfund/ sites/rods/fulltext/r0289097.pdf [accessed Jan. 9, 2007].

EPA (U.S. Environmental Protection Agency). 1989b. Risk Assessment Guidance for Superfund, Volume 1: Human Health Evaluation Manual (Part A, Interim Final). EPA/540/1-89/002. Office of Emergency and Remedial Response, U.S. Environmental Protection Agency, Washington, DC. December 1989 [online]. Available: http://rais.ornl.gov/homepage/HHEMA.pdf [accessed May 9, 2007].

EPA (U.S. Environmental Protection Agency). 1989c. Record of Decision: Commencement Bay, Near Shore/Tide Flats. EPA ID: WAD980726368 OU 01, 05, Pierce County, WA. EPA/ROD/R10-89/020. Superfund Information Systems, U.S. Environmental Protection Agency [online]. Available: http:// www.epa.gov/superfund/sites/rods/fulltext/r1089020.pdf. [accessed Jan. 9, 2007].

EPA (U.S. Environmental Protection Agency). 1990a. Record of Decision: New Bedford. EPA ID: MAD980731335, OU 02, New Bedford, MA. EPA/ROD/R01-90/045. Superfund Information Systems, U.S. Environmental Protection Agency [online]. Available: http://www.epa.gov/superfund/ sites/rods/fulltext/r0190045.pdf [accessed Jan. 9, 2007].

EPA (U.S. Environmental Protection Agency). 1990b. EPA Superfund Explanation of Significant Differences: Bayou Bonfouca. EPA ID: LAD980745632, OU 01, Slidell, LA. EPA/ESD/R06-90/900. Superfund Information Systems, U.S. Environmental Protection Agency [online]. Available: http://www.epa. gov/superfund/sites/rods/fulltext/e0690900.pdf [accessed Jan. 9, 2007].

EPA (U.S. Environmental Protection Agency). 1991. Record of Decision: General Motors (Central Foundry Division). EPA ID: NYD091972554. OU 01 Massena, NY. EPA/ROD/R02-91/131. Superfund Information Systems, U.S. Environmental Protection Agency [online]. Available: http://www.epa.gov/superfund/sites/rods/fulltext/r0291131.pdf [accessed Jan. 9, 2007].

EPA (U.S. Environmental Protection Agency). 1993a. Record of Decision: E.I. Du Pont De Nemours & Co., Inc. (Newport Pigment Plant Landfill). EPA ID: DED980555122, OU 01, Newport, DE. EPA/ROD/R03-93/170. Superfund Information Systems, U.S. Environmental Protection Agency [online]. Available: http://www.epa.gov/superfund/sites/rods/fulltext/r0393170.pdf [accessed Jan.9, 2007].

EPA (U.S. Environmental Protection Agency). 1993b. Record of Decision: Reynolds Metals Co. EPA ID: NYD002245967, OU 01, Massena, NY. EPA/ROD/R02-93/201. Superfund Information Systems, U.S. Environmental Protection Agency [online]. Available: http://www.epa.gov/superfund/sites/rods/fulltext/r0293201.pdf [accessed Jan. 9, 2007].

EPA (U.S. Environmental Protection Agency). 1994. PCB Action Level for Manistique River and Harbor. Letter to William E. Muno, Director, Waste Management Division, from Milton Clark, Senior Health and Science Advisor, U.S. Environmental Protection Agency, Chicago, IL. August 4, 1994.

EPA (U.S. Environmental Protection Agency). 1995. Superfund Site Close-out Report Marathon Battery Company Site Putnam County Cold Spring, New York. September 28, 1995.

EPA (U.S. Environmental Protection Agency). 1996a. Groundwater Remedial Action Five-Year Review: Bayou Bonfouca Superfund Site, Slidell, Louisiana. September 1996.

EPA (U.S. Environmental Protection Agency). 1996b. Record of Decision: United Heckathorn Co. EPA ID: CAD981436363, OU 01, Richmond, CA. EPA/ROD/R09-96/502. Superfund Information Systems, U.S. Enviromental Protection Agency [online]. Available: http://www.epa.gov/superfund/sites/rods/fulltext/r0996502.pdf [accessed Jan. 9, 2007].

EPA (U.S. Environmental Protection Agency). 1997a. Five-Year Review Type Ia Outboard Marine Corporation Site, Waukegan, Illinois. U.S. Environmental Protection Agency Region 5 [online]. Available: http://www.epa.gov/superfund/sites/fiveyear/f97-05002.pdf [accessed May 22, 2006].

EPA (U.S. Environmental Protection Agency). 1997b. Explanation of Significant Differences: Commencement Bay, Near Shore/Tide Flats. EPA ID: WAD980726368, OU 01, Pierce County, WA. EPA/ESD/R10-97/059. Superfund Information Systems, U.S. Environmental Protection Agency [online]. Available: http://www.epa.gov/superfund/sites/rods/fulltext/e1097059.pdf [accessed Jan. 9, 2007].

EPA (U.S. Environmental Protection Agency). 1997c. Record of Decision: Harbor Island (Lead). EPA ID: WAD980722839, OU 07, Seattle, WA. EPA/ROD/R10-97/045. Superfund Information Systems, U.S. Environ-

mental Protection Agency [online]. Available: http://www.epa.gov/super
fund/sites/rods/fulltext/r1097045.pdf [accessed Jan. 9, 2007].

EPA (U.S. Environmental Protection Agency). 1998a. Five-Year Review Report
Marathon Battery Company Superfund Site Village of Cold Spring,
Putnam County, New York. U.S. Environmental Protection Agency,
Region 2, New York. June 1998 [online]. Available: http://www.epa.gov/
superfund/sites/fiveyear/f98-02005.pdf [accessed May 4, 2007].

EPA (U.S. Environmental Protection Agency). 1998b. Record of Decision: New
Bedford Harbor OU, 01 Upper and Lower Harbor, New Bedford, MA
September 25, 1998. Superfund Information Systems, U.S. Environmental
Protection Agency [online]. Available: http://www.epa.gov/ne/superfund/
sites/newbedford/38206.pdf [accessed Jan.9, 2007].

EPA (U.S. Environmental Protection Agency). 1999. Evaluation of Toxicity and
Bioaccumulation of Contaminants in Sediments Samples from Waukegan
Harbor, Illinois. EPA-905-R-99-009. Great Lakes National Program Office,
U.S. Environmental Protection Agency, Chicago, IL. October 1999 [online].
Available: http://www.epa.gov/glnpo/sediment/waukegan/whrpt.pdf [ac-
cessed Jan. 9, 2007].

EPA (U.S. Environmental Protection Agency). 2000a. Realizing Remediation II: A
Summary of Contaminated Sediment Remediation Activities at Great
Lakes Areas of Concern. U.S. Environmental Protection Agency, Great
Lakes National Program Office, Chicago, IL. July 2000 [online]. Available:
http://www.epa.gov/glnpo/sediment/realizing2/index.html [accessed Aug.
16, 2006].

EPA (U.S. Environmental Protection Agency). 2000b. Record of Decision: Ketchikan Pulp Company. EPA ID: AKD009252230, OU02, Ketchikan, AK.
EPA/ROD/R10-00/035. Superfund Information Systems, U.S. Environmental Protection Agency [online]. Available: http://www.epa.gov/super
fund/sites/rods/fulltext/r1000035.pdf [accessed Jan. 10, 2007].

EPA (U.S. Environmental Protection Agency). 2000c. Record of Decision: Puget
Sound Naval Shipyard Complex,. EPA ID: WA2170023418, OU 02 Marine
Bremerton, WA. EPA/ROD/R10-00/516..June 13, 2000. Superfund Informa-
tion Systems, U.S. Environmental Protection Agency [online]. Available:
http://www.epa.gov/superfund/sites/rods/fulltext/r1000516.pdf   [accessed
May 4, 2007].

EPA. (U.S. Environmental Protection Agency). 2000d. Record of Decision:
Newport Naval Education and Training Center. EPA ID: RI6170085470,
OU4, Newport, RI. EPA/ROD/R01-00/155. Superfund Information Systems,
U.S. Environmental Protection Agency [online]. Available: http://www.epa.
gov/superfund/sites/rods/fulltext/r0100155.pdf [accessed Jan. 10, 2007].

EPA (U.S. Environmental Protection Agency). 2002. Second Five-Year Review
Report for Outboard Marine Corporation Superfund Site, Waukegan, Lake

Country, IL. U.S. Environmental Protection Agency, Chicago, IL. September 2002 [online]. Available: http://www.epa.gov/R5Super/fiveyear/ reviews_pdf/illinois/outboard_marine.pdf [accessed May 22, 2006].

EPA (U.S. Environmental Protection Agency). 2003a. Sediment Remedy Re-Evaluation: Bayou Bonfouca Site, Slidell, St. Tammany Parish, Louisiana. U.S. Environmental Protection Agency, Region 6. February 2003.

EPA (U.S. Environmental Protection Agency). 2003b. Five-Year Review Report Marathon Battery Company Superfund Site, Village of Cold Spring, Putnam County, New York. U.S. Environmental Protection Agency, Region 2, New York, NY. June 2003 [online]. Available: http://www.epa. gov/superfund/sites/fiveyear/f03-02019.pdf [accessed May 4, 2007].

EPA (U.S. Environmental Protection Agency). 2003c. Explanation of Significant Differences: Harbor Island (Lead). EPA ID: WAD980722839, OU 07, Seattle, WA. EPA/ESD/R10-03/011. Superfund Information Systems, U.S. Environmental Protection Agency [online]. Available: http://www.epa.gov/super fund/sites/rods/fulltext/e1003011.pdf [accessed Jan. 10, 2007].

EPA (U.S. Environmental Protection Agency). 2003d. Explanation of Significant Differences: Harbor Island (Lead). EPA ID: WAD980722839, OU 07, Seattle, WA. EPA/ESD/R10-03/010. Superfund Information Systems, U.S. Environmental Protection Agency [online]. Available: http://www.epa.gov/ superfund/sites/rods/fulltext/e1003010.pdf [accessed Jan. 10, 2007].

EPA (U.S. Environmental Protection Agency). 2005a. Five-Year Review Report: General Motors (Central Foundry Division) Superfund Site, St. Lawrence County, Town of Massena, New York. U.S. Environmental Protection Agency, Region 2, New York, NY. July 2005 [online]. Available: http:// www.epa.gov/superfund/sites/fiveyear/f05-02010.pdf [accessed Jan. 10, 2006].

EPA (U.S. Environmental Protection Agency). 2005b. Second Five-Year Review Report. E. I. du Pont de Nemours & Co., Inc. (Newport Pigment Plant Landfill) Superfund Site (a.k.a. DuPont-Newport Site), Newport, Delaware. U.S. Environmental Protection Agency, Region 3, Philadelphia, PA. March 31, 2005 [online]. Available: http://www.epa.gov/superfund/sites/fiveyear/ f0503006.pdf [accessed Jan. 10, 2007].

EPA (U.S. Environmental Protection Agency). 2005c. Second Five-Year Review Report: Harbor Island Superfund Site, Seattle, King County, Washington. U.S. Environmental Protection Agency, Region 10, Seattle, WA. September 2005 [online]. Available: http://yosemite.epa.gov/R10/CLEANUP.NSF/ 9f3c21896330b4898825687b007a0f33/5a64831b6521f46b8825650200836f1c/$F ILE/2nd%205%20Year%20Review%209-27-05.pdf [accessed Jan. 10, 2007].

EPA (U.S. Environmental Protection Agency). 2006a. Case Study Data and Information. U.S. Environmental Protection Agency.

EPA (U.S. Environmental Protection Agency). 2006b. Five-Year Review Report. Third Five-Year Review Report for Bayou Bonfouca Superfund Site, Slidell, St., Tammany Parish, Louisiana. U.S. Environmental Protection Agency, Region 6, Dallas, TX. May 2006 [online]. Available: http://epa.gov/earth1r6/6sf/pdffiles/bb_5yr_2006-0707.pdf [accessed Jan. 10, 2007].

EPA (U.S. Environmental Protection Agency). 2006c. Five-Year Review Report: Reynolds Metals Company Site, St. Lawrence County, Town of Massena, New York. U.S. Environmental Protection Agency, Region 2, New York, NY. April 2006 [online]. Available: http://www.epa.gov/superfund/sites/fiveyear/f06-02018.pdf [accessed Jan. 10, 2007].

EPA (U.S. Environmental Protection Agency). 2007a. Sediment Assessment and Remediation Report Assessment of Sediment Quality in the Black River Watershed. Final Report. U.S. Environmental Protection Agency, Great Lakes Contaminated Sediments [online]. Available: http://www.epa.gov/glnpo/sediment/BlackRiver/FinalReport.htm [accessed May 4, 2007].

EPA (U.S. Environmental Protection Agency). 2007b. Outboard Marine Corp. EPA ID: ILD000802827. NPL Fact Sheet. U.S. Environmental Protection Agency, Region 5 Superfund [online]. Available: http://www.epa.gov/region5/superfund/npl/illinois/ILD000802827.htm [accessed May 4, 2007].

Exponent. 2005. 2004 Monitoring Report for Sediment Remediation in Ward Cove, Alaska. Prepared for Ketchikan Pulp Company, Ketchikan, AK, by Exponent, Bellevue, WA. June 2005 [online]. Available: http://yosemite.epa.gov/R10/CLEANUP.NSF/9f3c21896330b4898825687b007a0f33/2dd5ab7462e4f004882567b30057eb7b/$FILE/KPC_monrpt_2004.pdf [accessed May 4, 2007].

Fort James Corporation, Foth &Van Dyke, and Hart Crowser Inc. 2001. Final Report 2000 Sediment Management Unit 56/57 Project, Lower Fox River, Green Bay, Wisconsin. Prepared for U.S. Environmental Protection Agency and Wisconsin Department of Natural Resources, by Fort James Corporation, Foth &Van Dyke, and Hart Crowser Inc. January 2001 [online]. Available: http://dnr.wi.gov/org/water/wm/foxriver/documents/finalreport/final_report.pdf [accessed Jan. 10, 2007].

Foth & Van Dyke. 2000. Summary Report: Fox River Deposit N. Wisconsin Department of Administration, Wisconsin Department of Natural Resources. April 2000 [online]. Available: http://dnr.wi.gov/org/water/wm/foxriver/documents/sediment/depositn_report.pdf [accessed Jan. 10, 2007].

Fox, R., M. Jury, and J. Kern. 2006. Overview of Lower Fox River OU1 Dredge Residuals. Presentation at the Second Meeting on Sediment Dredging at Superfund Megasites, June 7, 2006, Irvine, CA.

Fox River Group. 1999. Effectiveness of Sediment Removal. September 27, 1999 (Appendix C in GE 2000).

Goovaerts, P. 1997. Geostatistics for Natural Resources Evaluation. New York: Oxford University Press.

Green, G., V. Buhr, and M. Binsfeld. 2007. Controlling Dredge Cut Accuracy, Sediment Resuspension, and Sediment Dewatering: Fox River OU1 Contaminated Sediment Project. Presentation at the 4th International Conference on Remediation of Contaminated Sediments, January 22-25, 2007, Savannah, GA.

Helsel, D.R. 2005. Nondetects and Data Analysis: Statistics for Censored Environmental Data. Hoboken, NJ: Wiley. 250pp.

Herrenkohl, M., L. Jacobs, J. Lally, G. Hartman, B. Hogarty, K. Keeley, J. Sexton, and S. Becker. 2006. Ward Cove sediment remediation project revisited: Long-term success of thin-layer placement remedy. Pp. 421-431 in Proceedings of the Western Dredging Association 26th Technical Conference and 38th Annual Texas A&M Dredging Seminar, June 25-26, 2006, San Diego, CA, R.E. Randell, ed. Center for Dredging Studies, Ocean Engineering Program, Civil Engineering Department, Texas A&M Univer-sity, College Station, TX.

Hollander, M., and D. Wolfe. 1973. Nonparametric Statistical Inference. New York: Wiley.

ILDPH/ATSDR (Illinois Department of Public Health and the Agency for Toxic Substances and Disease Registry). 2004. Health Consultation. Outboard Marine Corporation/Waukegan Harbor Waukegan, Lake County, Illinois. EPA Facility ID: ILD000802827. Illinois Department of Public Health and the Agency for Toxic Substances and Disease Registry. April 20, 2004 [online]. Available: http://www.atsdr.cdc.gov/HAC/PHA/outboard2/omc_toc.html [accessed Jan. 10, 2006].

Kemble, N.E., D.G. Hardesty, C.G. Ingersoll, B.T. Johnson, F.J. Dwyer, and D.D. MacDonald. 2000. An evaluation of the toxicity of contaminated sediments from Waukegan Harbor, Illinois, following remediation. Arch. Environ. Contam. Toxicol. 39(4):452-461.

King County. 1999. King County Combined Sewer Overflow Water Quality Assessment for the Duwamish River and Elliott Bay. King County Department of Natural Resources, Seattle WA; and Parametrix, Inc. Kirkland, WA. February 1999 [online]. Available: http://dnr.metrokc.gov/wlr/waterres/wqa/wqrep.htm [accessed April 25, 2007].

Levinton, J.S., S.T. Pochron, and M.W. Kane. 2006. Superfund dredging restoration results in widespread regional reduction in cadmium in blue crabs. Environ. Sci. Technol. 40(24):7597-7601.

Mackie, J.A., S.M. Natali, J.S. Levinton, and S.A. Sanudo-Wilhelmy. 2007. Declining metal levels at Foundry Cove (Hudson River, New York): Response to localized dredging of contaminated sediments. Environ. Pollut. 149(2):141-148.

Malcolm Pirnie, Inc. and TAMS Consultants, Inc. 2004. Final Engineering Performance Standards Hudson River PCBs Superfund Site, Appendix: Case Studies of Environmental Dredging Projects. Prepared for U.S. Army Corps of Engineers, Kansas City District, by Malcolm Pirnie, Inc., White Plains, NY, and TAMS Consultants, Inc., Bloomfield, New Jersey. April 2004 [online]. Available: http://www.epa.gov/hudson/eng_perf/FP5001.pdf [accessed Aug. 16, 2006].

MIDEQ (Michigan Department of Environmental Qualilty). 1996. Manistique River Area of Concern Manistique, Michigan Remedial Action Plan Update. February 20, 1996 [online] Available: http://www.epa.gov/glnpo/aoc/mantique/1996_Manistique_RAP.pdf. [accessed May 4, 2007].

Montgomery Watson. 2001. Final Summary Report: Sediment Management Unit 56/57 Demonstration Project, Fox River, Green Bay, Wisconsin. Prepared for Fox River Group and Wisconsin Department of Natural Resources, By Montgomery Watson. September 2001 [online]. Available: http://dnr.wi.gov/org/water/wm/foxriver/documents/finalreport/final_summary_report.pdf [accessed Jan. 10, 2007].

Nadeau, S. 2006. SMWG Review and Analysis of Selected Sediment Dredging Projects. Presentation at the Second Meeting on Sediment Dredging at Superfund Megasites, June 7, 2006, Irvine, CA.

Nadeau, S.C., and M.M. Skaggs. 2007. Analysis of Recontamination of Completed Sediment Remedial Projects. Paper D-050 in Remediation of Contaminated Sediments-2007: Proceedings of the 4th International Conference on Remediation of Contaminated Sediments, January 2007, Savannah, GA. Columbia, OH: Battelle Press.

NRC (National Research Council). 2001. A Risk-Management Strategy for PCB-Contaminated Sediments. Washington, DC: National Academy Press.

NYSDEC (New York State Department of Environmental Conservation). 1997. Record of Decision: Cumberland Bay Sludge Bed-Wilcox Dock Site (OU-1), Plattsburgh, Clinton County Site Number 5-10-017. New York State Department of Environmental Conservation, Division of Environmental Remediation. December 1997.

NYSDEC (New York State Department of Environmental Conservation). 2001. Cumberland Bay Sludge Bed Removal Project. New York State Department of Environmental Conservation. April 2001.

Ortiz, E., R.G. Luthy, D.A. Dzombak, and J.R. Smith. 2004. Release of polychlorinated biphenyls from river sediment to water under low-flow conditions: Laboratory assessment. J. Environ. Engin. 130(2):126-135.

Patmont, C. 2006. Contaminated Sediment Dredging Residuals: Recent Monitoring Data and Management Implications. Presentation at the Second Meeting on Sediment Dredging at Superfund Megasites, June 7, 2006, Irvine, CA.

Patmont, C., and M. Palermo. 2007. Case Studies Environmental Dredging Residuals and Management Implications. Paper D-066 in Remediation of Contaminated Sediments-2007: Proceedings of the 4th International Conference on Remediation of Contaminated Sediments, January 2007, Savannah, GA. Columbia, OH: Battelle Press.

Qian, P.Y, J.W. Qiu, R. Kennish, and C.A. Reid. 2003. Recolonization of benthic infauna subsequent to capping of contaminated dredged material in East Sha Chau, Hong Kong. Estuar. Coast. Shelf S. 56(3-4):819-831.

Reible, D.D, , D. Hayes, C. Lie-Hing , J. Patterson, N. Bhowmik, M. Johnson, and J. Teal. 2003. Comparison of the long-term risks of removal and in-situ management of contaminated sediments in the Fox River. Soil Sediment Contam. 12(3):325-344.

Stern, J., and C.R. Patmont. 2006. Evaluation of Post-Dredge Monitoring Results to Assess Net Risk Reduction of Different Sediment Cleanup Options. Presentation at the SETAC North America 27th Annual Meeting, Nov. 5-9, 2006, Montreal, Canada.

Stern, J., C.R. Patmont, and D. Hennessy. 2007. Evaluation of Post-dredge Monitoring Results to Assess Net Risk Reduction of Different Sediment Cleanup Options. Presentation at 2007 Georgia Basin Puget Sound Research Conference, March 26–29, 2007, Vancouver, BC.

Steuer, J.J. 2000. A Mass-Balance Approach for Assessing PCB Movement During Remediation of a PCB-Contaminated Deposit on the Fox River, Wisconsin. U.S. Geological Survey Water Resources Investigation Report 00-4245. U.S. Department of the Interior, U.S. Geological Survey. December 2000 [online]. Available: http://www.dnr.state.wi.us/ORG/water/wm/foxriver/cuments/sediment/usgs_5657_report.pdf [accessed Jan. 11, 2007].

Stow, C.A., S.R. Carpenter, L.A. Eby, J.F. Amrhein, and R J Hesselberg 1995. Evidence that PCBs are approaching stable concentrations in Lake Michigan fishes. Ecol. Appl. 5(1):248-260.

TetraTech. 2004. Final Interim Remedial Action Report for Site 1—McAllister Point Landfill, Operable Unit 4—Marine Sediment/Management of Migration, Naval Station Newport, Middletown, Rhode Island. Prepared for Department of the Navy, Engineering Field Activity, Northeast Naval Facilities Engineering Command, Lester, PA, by Tetra Tech. FW, ICN, Langhorne, PA. September 28, 2004.

TetraTech NUS. 2006. Round 1: December 2004 Long-Term Monitoring Report for McAllister Point Landfill Naval Station Newport Newport, Rhode Island. Prepared for Engineering Field Activity Northeast Naval Facilities Engineering Command, Lester, PA. Contract Number N62472-03-D-0057. March 2006.

Thibodeaux, L.J., and K.T. Duckworth. 1999. The effectiveness of environmental dredging: A study of three sites. Remediat. J. 11(3):5-33.

URS (URS Woodward Clyde Group Consultants Inc.). 1999. Construction Moni-
    toring Report Christina River Remediation: Newport Superfund Site,
    Newport, DE. Prepared for DuPont Corporate Remediation Group, Wil-
    mington, Delaware, by URS Woodward Clyde Group Consultants Inc.,
    Buffalo, New York. December 1999.
URS (URS Group Inc.). 2006. Final 2003 Marine Monitoring Report, OU B Marine,
    Bremerton Naval Complex, Bremerton, Washington. Prepared for Naval
    Facilities Engineering Command Northwest, Poulsbo, WA, by URS Group,
    Inc., Seattle, WA. February, 2, 2006.
USACE (U.S. Army Corps of Engineers). 1992. Remediation of Marathon Battery
    Superfund Project: Cold Spring, New York, Supplemental Information. IFB
    No. DACW41-92-B-0030. U.S. Army Corps of Engineers, Kansas City Dis-
    trict. May 1992.
USACE (U.S. Army Corps of Engineers). 1995. New Bedford Hot Spot after
    Dredge Sediment Sampling. Memorandum to Dave Dickerson, U.S. Envi-
    ronmental Protection Agency, from Mark Otis, U.S. Army Corps of Engi-
    neers. CENED-PD-E. April 19, 1995.
USACE (U.S. Army Corps of Engineers). 2005. Benthic Recovery Review. Draft
    Technical Note ERDC TN-DOER-E. August 2005.
Weston (Weston Solutions Inc.). 2002. Final Comprehensive Post-Removal Sum-
    mary Report Manistique Harbor and River Site U.S. Route 2 Highway
    Bridge Manistique, Michigan. Prepared for the U.S. Environmental Protec-
    tion Agency, Region 5. November 12, 2002.
Weston (Weston Solutions Inc.). 2005a. Final Manistique Harbor and River Site,
    Manistique, Michigan, Data Evaluation Report: Revision 1-19 May 2005.
    Document No. RWW236-2A-ATGN. Prepared for U.S. Environmental Pro-
    tection Agency, Chicago, IL.
Weston (Weston Solutions Inc.). 2005b. Human Health Risk Assessment
    GE/Housatonic River Site, Rest of River, Vol. 1. DCN:GE-021105-ACMT.
    Environmental Remediation Contract, GE/Housatonic River Project, Pitts-
    field, MA. Prepared for U.S. Army Corps of Engineers, New England Dis-
    trict, Concord, MA, and U.S. Environmental Protection Agency, Boston,
    MA, by Weston Solutions Inc., West Chester, PA. February 2005 [online].
    Available: http://www.epa.gov/NE/ge/thesite/restofriver/reports/hhra_219
    190/219190_HHRA_Vol1.pdf [accessed May 7, 2007].
WI DNR/EPA (Wisconsin Department of Natural Resources and the U.S. Envi-
    ronmental Protection Agency). 2002. Record of Decision, Operable Unit 1
    and Operable Unit 2, Lower Fox River and Green Bay. Wisconsin Depart-
    ment of Natural Resources, Madison, WI, and the U.S. Environmental Pro-
    tection Agency, Region 5, Chicago, IL. December 2002 [online]. Available:
    http://dnr.wi.gov/org/water/wm/foxriver/documents/whitepapers/Final_R
    ecord_of_Decision_OUs_1-2.pdf [accessed Jan. 11, 2007].

Windward. 2005. Lower Duwamish Waterway Remedial Investigation. Data
    Report: Fish and Crab Tissue Collection and Chemical Analyses. Final Re-
    port. Prepared for U.S. Environmental Protection Agency, Region 10, Seat-
    tle WA, and Washington State Department of Ecology, Bellevue, WA, by
    Windward Environmental LLC, Seattle, WA. July 27, 2005 [online]. Avail-
    able: http://www.ldwg.org/assets/fish_crab_tissue/final_fish%2bcrab_
    tissue_dr.pdf [accessed April 25, 2007].
Windward. 2006. Lower Duwamish Waterway Remedial Investigation. Data
    Report: Chemical Analyses of Fish and Crab Tissue Samples Collected in
    2005. Final Report. Prepared for U.S. Environmental Protection Agency,
    Region 10, Seattle WA, and Washington State Department of Ecology,
    Bellevue, WA, by Windward Environmental LLC, Seattle, WA. April 19,
    2006 [online]. Available: http://www.ldwg.org/assets/fish_crab_tissue/2005
    dr/2005_fish%2bcrab_dr.pdf [accessed April 25, 2007].
Zarull, M.A., J.H. Hartig, and L. Maynard. 1999. Ecological Benefits of Contami-
    nated Sediment Remediation in the Great Lakes Basin. Windsor, Ontario:
    International Joint Commission [online]. Available: http://www.ijc.org/
    php/publications/html/ecolsed/ [accessed Sept. 12, 2006].

5

Monitoring for Effectiveness: Current Practices and Proposed Improvements

MONITORING FOR EFFECTIVENESS

The effectiveness of environmental dredging in reducing risk, as predicted when the remedy was selected, can be verified only through monitoring. Monitoring includes

- Monitoring of potential short-term risks due to dredging.
- Verification that dredging has achieved its immediate target cleanup levels.
- Long-term monitoring to determine whether remedial objectives have been or are likely to be achieved in the expected time frame.

Monitoring of effectiveness is an essential part of the remedy and should be proportional to the size of the project. Through careful monitoring it is possible to demonstrate whether environmental dredging minimizes risks to human and ecologic receptors during active operations and to judge the success of contaminant cleanup in decreasing risk after the cessation of active remedial operations. Monitoring is the only way to determine short-term and long-term compliance with remedial-action objectives and evaluate net risk reduction of the remediation, and it forms the basis of the 5-year performance reviews after cleanup. Because sediments typically pose long-term risks, monitoring often must span decades to assess risk reduction.

The ultimate goal of monitoring is protection—that is, ensuring that short-term and long-term risks are minimized, by providing sufficient information to judge that the remedy is effective, or to adapt site management to optimize the remedy's performance to achieve risk-based objectives. Management adaptation may entail modification of dredging procedures—for example, if short-term exposures exceed expected magnitudes—or modification of the remedy itself by amendment or modification of the record of decision (ROD) if long-term risk reduction proceeds more slowly or more rapidly than expected.

An effective sediment-monitoring plan takes into account the successive stages of sediment cleanup: site characterization; selection, planning, and implementation of the remedial action; effectiveness assessment; and adaptive management.[1] Monitoring should build on the studies previously performed for the remedial investigation and feasibility study (RI/FS), which should have

- Determined the nature and extent of contamination and any trends in time (for example, due to natural recovery).
- Supported or developed a conceptual site model.
- Provided information to assess risks to the environment and people.
- Evaluated remedial alternatives, including a quantitative comparison of risks associated with implementation of each one.

Once the remedy is selected and implementation begins, monitoring extends the record of site conditions into the future.

MONITORING PRINCIPLES

1. Monitoring should be based on and inform the conceptual site model.

---

[1]In general, adaptive management is the testing of hypotheses and conclusions and re-evaluation of site assumptions and decisions as new information is gathered (see Chapter 6 for further discussion). It is an important component of the updating of the conceptual site model (EPA 2005a).

a. Appropriate metrics need to be chosen, measuring success against expectations based on the conceptual site model.

b. Monitoring is an essential verification step, not an add-on activity or a second remedial investigation.

2. Effective monitoring of the remedy requires characterization of pre-remedial trends and reference conditions, in addition to post-remedial trends.

a. Sufficient pre- and post-remedial sample sizes are needed, to allow for natural heterogeneity.

b. The time span of pre- and post-remedial sampling needs to be sufficient to capture the time scale of recovery processes.

c. Proper reference sites and conditions must be specified and monitored.

Monitoring and the Conceptual Site Model

Links between contamination, exposure, and risk can be highly complex, involving multiple physical, chemical, and biologic processes. A particular combination of these is present at each site. Monitoring protocols and media to be monitored will vary accordingly, and should be closely linked to site conceptual models that link site conditions with biologic exposures and effects (EPA 1998). The expectations of the Superfund ROD are a natural yardstick against which to judge effectiveness. Those expectations of short-term exposures and long-term risk reduction due to dredging should be based on the conceptual site model and its mathematical counterpart.

Where site conceptual models are insufficiently developed, it is difficult to develop an understanding of the factors driving trends in site-monitoring data. On major dredging sites, short-term and long-term expectations based on site models will have been developed as part of the feasibility study supporting remedy selection. Collecting data to test whether expectations have been fulfilled is part of the process of conceptual-model development, testing, and refinement that was begun with the initial site characterization. If the important cause-effect relationships between contaminant sources, transport mechanisms, exposure pathways, and receptors have been well characterized by the time a remedy is selected, including bioavailability and food-web relationships as applicable, and there has been sufficient pilot testing or other means of anticipating site-specific field conditions and implementation challenges, well-designed monitoring should indicate the remedy has performed as expected. If not, monitoring can help to identify important elements that are missing from the conceptual site model so that its predictions can be made more accurate and site management can be adapted accordingly, as recommended in EPA's *Contaminated Sediment Remediation Guidance* (2005a).

In monitoring of the effectiveness of a remedy, important transport mechanisms and exposure pathways to be monitored include not only the ones that control exposures and risks under normal conditions, but also the ones that may be triggered by dredging, such as releases that may occur during normal dredging operations or when debris or bedrock is encountered. Therefore, before selection and implementation of a remedy, the site investigation should thoroughly examine factors that would complicate dredging and include them in the conceptual model. Complicating site conditions and operational limitations can also be identified through pilot studies to verify the performance of the selected technology under site-specific conditions.

Data collection is one of the more expensive aspects of site management (Box 5-1).[2] Judicious use of the conceptual site model in designing the monitoring plan focuses data collection where it can best ensure protectiveness while conserving monitoring resources. Monitoring should target the key pathways and receptors necessary to determine whether remedial objectives have been met. If dredging is intended to reduce ecologic or human health risks, the conceptual site model can be used to focus sampling on locations and receptors that directly indicate risk related to the targeted sediments and contaminants and minimize spurious effects, such as increased body burdens in migratory species

---

[2]In addition to the example provided in Box 5-1, see the breakdown of costs of the Hylebos Waterway and 2004 dredging at New Bedford Harbor presented in Chapter 2 (Figures 2-4 and 2-5). However, it should be noted that these costs may not be directly comparable; it is not clear, for example, whether the costs include design costs and long-term monitoring.

BOX 5-1 Estimated Monitoring Costs for Lower Fox River and
Green Bay, Wisconsin, ROD Remedy

The costs of construction monitoring (including verification sampling) and
long-term monitoring (including an initial pre-dredging baseline survey of affected media and surveys of the same media continuing for decades after the
remedy) for the ROD remedy for Operable Units 2-5 of the Lower Fox River and
Green Bay, WI, have been estimated at $6 and $8 million, respectively (Shaw
2006). Together those costs exceed the estimated cost of engineering and construction support for the remedy, including development of design documentation, plans, and specifications.

and species with wide home ranges that are due to unrelated exposures
at remote locations. If dredging is intended to minimize water-column
contaminant transport, the site model can be used to control for the effects of flow, temperature, seasonality, non-sediment-related stressors
(such as point and nonpoint sources), and other ambient conditions to
inform sampling plans and assist in interpreting the results.

Monitoring decisions may be influenced by financial, jurisdictional,
or political interests, even though they should be guided solely by the
need to verify conceptual site models, inform remedy implementation,
and to document when remedial objectives have been achieved. Cleanup
negotiations between regulators and responsible parties can be contentious, and agreements on the scope of cleanups are often the results of a
long and difficult process. The scope of post-remedial monitoring can
also be established during those negotiations. The parties have few incentives to seek actively to establish whether a chosen remedial action
had its intended effect. This paradigm, wherein both regulators and responsible parties may perceive that they have something to lose and
nothing to gain in a robust post-remediation monitoring program, may
be a reason for the lack of post-remediation confirmation sampling seen
at some sites. Public-sector and private-sector designers of a monitoring
plan may face strong pressures to demonstrate early success while controlling costs and may also feel pressure to divert remedial funding to
support broader long-term natural-resource monitoring efforts. Those
ancillary goals may be attractive to parties involved in designing a monitoring program, but the fundamental objectives of monitoring are to per-

form a fair and conclusive evaluation of remedy effectiveness and risk reduction, and resources and energy should be focused on this objective. Information developed from the monitoring program should be used to guide future decision-making in a manner which balances a realistic assessment of the projected environmental benefit relative to anticipated costs.

Developing a body of well-designed site evaluations of dredging effectiveness will meet the broader programmatic objective of providing EPA and other lead agencies with invaluable information on strengths and weaknesses of dredging as a remedy—information that they can use in future remedial decision-making.

Comparisons to Baseline Conditions

To assess the effectiveness of the remedy, post-remedial monitoring should be compared with data trends and model forecasts developed before remedy selection. This requires that there be comparable datasets before (a "baseline") and after dredging. As stated by EPA (2005a, page 8-2),

> During site characterization, the project manager should anticipate expected post-remedy monitoring needs to ensure that adequate baseline data are collected to allow comparison of future datasets. Monitoring plans should also be designed to allow comparison of results with model predictions that supported remedy selection.

It is often difficult in practice for an effective monitoring plan to meet the above objectives. One important issue at Superfund megasites is that the time from initial site investigation to implementation of remedial measures can be 10 years or more; it is extremely difficult to ensure temporal and spatial consistency of baseline and post-remedial monitoring data, including data- quality assurance and control. Data collections that span many years can greatly complicate the selection of appropriate statistical tests for evaluating them. Those concerns are often manifested after the fact rather than being evident during the planning of the baseline and long-term monitoring programs.

Consistent with its role in supporting hypothesis-testing, the monitoring protocol should be rigorous enough to allow managers to evaluate critically the potential adverse effects of dredging on human and ecologic receptors and potential risk reductions due to removal of contaminated sediment. For example, proper reference sites or reference conditions should be established to allow comparison of affected media with pre-dredging or nondredged controls. Appropriate sample sizes should be determined from estimates of variability derived from pilot studies or other sources of data. In particular, the natural heterogeneity of biologic systems can be substantial and should be explicitly accounted for in defining sample sizes.

CURRENT MONITORING PRACTICES

According to Elzinga et al. (1998, as referenced in EPA 2004), monitoring is "the collection and analysis of repeated observations or measurements to evaluate changes in condition and progress toward meeting a management objective." Monitoring at Superfund sites is typically directed toward evaluation of the performance of a remedy and whatever environmental protections are in place during implementation of the remedy. Monitoring may include the collection of samples or real-time metered data

- During implementation of the remedy to assess immediate human health or environmental effects.
- Soon after implementation to determine compliance with cleanup levels or other short-term objectives.
- Over time to evaluate the achievement of the long-term remedial-action objectives, the need for maintenance or repair, and the continued effectiveness of the remedy and associated source control.

Ideally, the monitoring parameters measured are linked to site-specific risk factors so that success (or lack of success) of the remedy is evident and directly informs management of the site. There are no absolute requirements for monitoring elements or techniques, but a number of guidance documents have been published (Fredette et al. 1990; EPA/USACE 1998; EPA 2001a, 2004, 2005a) to identify relevant measurements and techniques and to guide the design of monitoring programs for a contaminated sediment site undergoing remediation.

Monitoring Parameters and Techniques

Monitoring involves combinations of physical, chemical, and biologic methods. Three critical lines of evidence that increasingly define successful sediment remediation include sediment physical stability, sediment chemical stability (lack of movement of contaminants from the sediment to the water column), and biologic-ecologic integrity. These three concepts are integral components of remedy evaluation, and monitoring should use techniques sufficient to measure progress toward these end points.

A variety of techniques and measurement parameters exist for the characterization of the nature, extent, and potential effects of sediments. These techniques range from relatively simple and quick to elaborate and time consuming (e.g., EPA 2001a; Wenning et al. 2005). Several of the techniques are described below and summarized in Box 5-2.

Physical Techniques

Available physical techniques include direct sampling of sediment for laboratory analysis of geophysical properties, core sampling to identify sediment layering or the presence of debris, side scan sonar to develop high resolution maps of bottom contours, acoustic sub-bottom profilers or magnetometers to map sub-bottom characteristics, remote sensing to document vegetative cover or other characteristics, videography or photography to document bottom features or shallow sediment profile characteristics, and instrumentation to measure environmental conditions (such as temperature and turbidity) or flow characteristics that may affect sediment and suspended solids transport. For example, sediment-profile imaging (a photographic technique) of surface (10-20 cm) characteristics can be conducted to establish various parameters including the depth of bioturbation, the depth of an oxygenated layer, general benthic community type and degree of recovery, or hydrogen sulfide gas production (see Figure 5-1 for an example). Other remote

BOX 5-2 Common Physical, Chemical, and Biologic Measurements Used
To Characterize Contaminated Sediments

Common physical measurements include

• Sediment geophysical properties, such as bulk density, particle size, and shear strength.
• Pre-dredging and post-dredging bottom elevations, and sediment bed-forms.
• Sediment layering, such as depth of disturbance or bioturbation, presence of gas bubbles, redox layers, and interfaces between sediment of different textures.
• Debris-field mapping (location, density, and size).
• Conductivity, temperature, turbidity, and suspended particles under various flow conditions.
• Stream velocities.

Common chemical measurements include

• Water-column parameters (such as dissolved oxygen and total and dissolved chemicals under various flow conditions).
• Surface- and subsurface -sediment chemistry, including magnitude, distribution, and depth of contamination.
• Pore water contaminant concentrations.
• Bioavailable fractions of contaminants in sediment, on the basis of organic-carbon normalization or acid volatile sulfide (AVS) analysis.
• Tissue contaminant concentrations including tissues ingested by humans (in field collected or exposed aquatic organisms or plants) or tissue surrogates.
• Air quality (including odor) during construction of remedy or handling of dredged material.

Common biologic measurements include

• Benthic invertebrate community structure (including abundance, diversity, and other structural or functional indexes).
• Toxicity (acute and chronic effects measured in the laboratory or field).
• Aquatic or wetland plant community structure (including species composition and percentage of cover).
• Fisheries status (including size, abundance, reproductive status, and incidence of lesions or parasites).

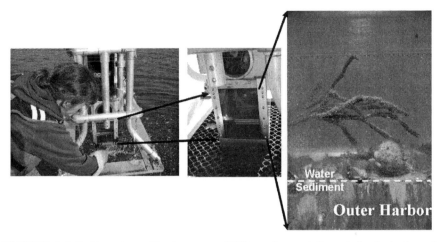

FIGURE 5-1 Sediment profile imagery (SPI) equipment (two left photos) and sediment profile photograph (right) from New Bedford Harbor Superfund site (the outer harbor is the area of the site with the least contamination). This equipment is used as part of the long-term monitoring program at the site to assess benthic quality rapidly and augment traditional benthic survey techniques that entail sieving and enumeration. Source: W. Nelson, U.S. Environmental Protection Agency.

sensing techniques, such as lidar (light detection and ranging) can be used to map large-scale site characteristics, including the extent of eel grass beds or other vegetative cover. Assessments of physical stability of sediments (which translates into the likelihood for sediments to be dislodged and transported by erosive events) are based on site uses, hydrology and geomorphology, sediment bed descriptions (radio dating deposits, stratigraphy, and physical characteristics), and measurement of sediment transport and sediment bed dynamics (erodability or bed elevation changes) (Bohlen and Erickson 2006).

Chemical Monitoring

Chemical monitoring can address multiple media—including air, sediment, water, biota, groundwater, and pore water—and can be designed to evaluate specific phases of chemicals of concern (for example, if they are dissolved or suspended in association with solids). It is important to monitor those parameters that affect chemical bioavailability, such as total and dissolved organic carbon, acid volatile sulfides (AVSs), grain size, and pore water fractions because organisms are exposed only to the bioavailable fraction (NRC 2003). The relationship between chemical concentrations, the bioavailable fraction, and toxic effects is the foundation for establishing sediment quality guidelines (see next section). Chemical sampling may involve in situ instrumentation for water, single-point grab samples of water or sediment obtained with various devices, or use of samplers that integrate chemistry over time or space (such as sediment traps, composite water samplers, and peepers). Rapid chemical screening techniques that use immunoassay response (enzyme-linked immunosorbent assays [ELISA]) or chemical fluorescence to document relative exposures have also been developed, but these are generally single-contaminant or contaminant-class tests, and few rapid field screening techniques are available for measuring a broad array of contaminants.

Some analytic methods for environmental samples can be time-consuming, labor-intensive, and expensive. For example, chemical measurements for persistent organic contaminants in sediments—such as polychlorinated biphenyls (PCBs), polycyclic aromatic hydrocarbons (PAHs), and DDT—require extraction, cleanup, and instrument analyses with gas chromatography or mass spectrometry. None of those measurements is rapid or performed conveniently in the field. As with sediments, the chemical analysis of biologic samples requires extraction, cleanup, and instrument techniques. Replicate measurements are necessary for both sediment samples and biologic tests because of inherent variability.

Newer techniques have been developed for deployment of manufactured materials in the form of passive sampling devices (such as semipermeable-membrane devices, solid-phase microextraction fibers, and Tenax) that can mimic biologic exposure to and tissue uptake of contaminants from water, pore water, or sediments. For example, semipermeable membrane devices (SPMDs) have been widely used in environmental applications since the early 1990s (Huckins et al. 1990; 1993) and applied at Superfund sites to monitor dissolved hydrophobic contaminants and estimate water column concentrations of these contaminants (e.g., Hofelt and Shea 1997; Weston 2005). Polyethylene devices (PEDs) passively sample hydrophobic organic compounds in the aqueous phase.

They are robust, simple, and inexpensive and have a short equilibration time (Booij et al. 1998; Adams 2003). The laboratory analysis of PEDs has fewer background-signal problems than the analysis of biologic samples. Another example is solid-phase microextraction (SPME) of sediment pore water, which may be quicker and more economical than conventional sampling. SPME uses fibers coated with a liquid polymer, a solid sorbent, or a combination. The fiber coating removes the compound from solution by sorption. The SPME fiber is then inserted directly into a gas chromatograph for desorption and analysis.

Some of these passive sampling devices can reach equilibrium with environmental conditions much more rapidly than living organisms. They have been widely used in recent years for sampling metals and organics in aquatic systems and found to be good indicators of fish and invertebrate bioaccumulation, that is, to be biomimetic (e.g., Arthur and Pawliszyn 1990; Huckins et al. 1993; Wells and Lanno 2001). Biomimetic samplers are easy to deploy and analyze, and they indicate exposure over time, but they are selective and do not indicate all chemical exposures or biologic effects (Table 5-1). Several of these techniques are still undergoing development and research is being conducted to better understand the relationship between the sampler concentrations, environmental concentrations, and bioaccumulation in organisms (e.g., Leslie et al. 2002; Lohmann et al. 2004; Vinturella et al. 2004; Conder and LaPoint 2005; You et al. 2006). With time and refinement, these technologies will likely become more available for routine application at contaminated sediment sites.

Biologic Monitoring

Biologic monitoring looks at the sublethal to lethal responses of individual organisms, populations, or communities in the environment or under controlled laboratory conditions. Biologic measurements and end points are usually more complex or difficult to obtain than physical or chemical measures, but biologic monitoring is the most definitive way to determine risk. Biologic monitoring typically provides a more integrated measurement of exposure (of both human and ecologic receptors) and is related more directly to ecosystem effects than is physical or chemical monitoring.

TABLE 5-1 Strengths and Limitations of Methods for Assessing Biologic Effects in Aquatic Ecosystems

Effect Assessment Method	Advantages	Limitations
Criteria or guidelines	Proven utility and ease of use	Assume single chemical effect; based on laboratory exposures; causality link uncertain
Biotic-ligand model	Proven utility and ease of use for accounting for metal bioavailability in surface water	Insufficient research and validation for use with sediment
Empirically based guidelines	Proven utility and ease of use	Bioavailability not accounted for; may lead to incorrect conclusion of presence or absence of risk
Equilibrium-based guidelines	Regulatory support; predictive capability	Not applicable in dynamic systems; does not consider all critical binding phases
Species sensitivity distributions	Use of all available data for derivation of EQC or PNEC	Lack of sufficiently large and diverse sediment-toxicity datasets
Indigenous biota	Target receptors; lack of laboratory extrapolation; long-term measure; proven utility; public interest; colonization and transplant methods increase stressor diagnostic power and experimental power	Habitat and other natural stressors or linkages confound causality linkage; inherent variability; loss of colonization units possible because of flow and vandalism
Tissue residues and biomarkers	Documents exposure; use for food web and risk models; widely used; very sensitive and timely	Adaptation, acclimation, and metabolism confound interpretations; uncertain adverse- effect threshold levels
Biomimetic devices: semipermeable membrane devices;	Accumulates organics or metals from waters and sediments through diffusion and sorption; amounts	Selectivity varies with different chemicals; may not mimic bioaccumulation of all organisms; some are subject

TABLE 5-1 Continued

Effect Assessment Method	Advantages	Limitations
solid-phase microextraction; Tenax; diffusive gradient transport	accumulated on these inert materials are similar to amounts bioaccumulated in fish tissues; can be placed in situ for short to long periods and then directly analyzed in laboratory	to fouling, depending on ecosystem; not standardized
Toxicity assays (laboratory)	Bioavailability indicator; proven utility; integrates effects of multiple chemicals; does not measure natural stressors	Causality link uncertain; laboratory-to-field extrapolation; individual-to-community extrapolations; does not measure natural stressors; cost of chronic assays
Toxicity and bioaccumulation assays (field)	More realistic exposure, which reduces artifact potential; measures many natural stressors and interactions; compartmentalizes exposures to various media; exposure-to-effect linkage is strong	Most methods are not standardized; limited use; deployment can be difficult; possible caging effects with some organisms; causality link uncertain; loss of units possible because of predators and vandalism; acclimation stress possible because of temperature, salinity, or hardness differences
Toxicity fractionation (laboratory)	Better establishes specific chemical causality; standard method for effluents	Subject to manipulation artifacts; acute toxicity only; limited use in sediments; large pore water volume requirements; limited sensitivity
Toxicity fractionation (field)	More realistic exposure, which reduces artifact potential; better establishes chemical causality	Very limited use; deployment can be difficult; shallow environments only; acute toxicity only; loss of units possible because of high flow and vandalism; not standardized

Source: Modified from Burton et al. 2005.

Biologic testing often has both field and laboratory components—organisms collected from the field are identified and enumerated or, in marine systems, exposed to sediment-bound or water-borne chemicals in a laboratory (e.g., Barbour et al. 1999; EPA 2001a; Adams et al. 2005). Indigenous organisms can be collected with nets, hooks, traps, grab samplers, or other devices. Standardized laboratory sediment toxicity and bioaccumulation testing methods are commonly used in assessments of the potential hazard of dredged materials (e.g., Environment Canada 1992; EPA 1994, 2000a; EPA/USACE 1998; ASTM 2006). Standardized methods are available for freshwater and marine systems, in both short and long term (chronic) exposures. These tests often are one component of a "Sediment Quality Triad" and other weight-of-evidence based approaches (Adams et al. 2005). A strong relationship has been documented between the responses in these standardized laboratory test responses (and indigenous benthic communities) and empirically-based sediment quality guidelines (discussed below) (Ingersoll et al. 2005).

Field-collected or laboratory-reared organisms can be deployed in cages or nets for defined periods of exposure to water or sediment and retrieved for analysis (e.g., Ireland et al. 1996; Tucker and Burton 1999; Burton et al. 2000; Chappie and Burton 2000; Greenberg et al. 2002; Adams et al. 2005; Crane et al. 2007). These caged-organism assays allow measurements of effects on growth and survival to be closely linked to environmentally-relevant chemical exposures (Table 5-1) (Solomon et al. 1997; Burton and Pitt 2002; Adams et al. 2005; Wharfe et al. 2007). Transplantation and recolonization of benthic macroinvertebrates on reference and site sediments have also been shown to be effective ways to measure site effects and risk, but they require exposures of up to a month (e.g., Clements and Newman 2002; Clark and Clements 2006). Biologic monitoring can be cost-effective, relative to chemical monitoring (Karr 1993; Hart 1994).

Transplantation, colonization, and caged-biota tests can be long and have deployment challenges. However, caged exposures often take only one to several days in freshwater systems (Ireland et al. 1996; Tucker and Burton 1999; Chappie and Burton 2000; Burton et al. 2002, 2005; Greenberg et al. 2002; Burton and Nordstrom 2004). In situ caged exposures of 2-4 days have been shown to provide uptake and toxicity information that is comparable with that of standardized laboratory tests that take 10-65 days at PCB-, chlorobenzene-, and metal-contaminated sediment sites (Greenberg et al. 2002; Burton et al. 2005). It is also useful to conduct laboratory-based exposures following standardized toxicity-test methods (EPA 2001a).

Monitoring Human Exposures

The biologic monitoring techniques listed and discussed above are related primarily to ecologic receptors and do not include monitoring of human subjects, which EPA and other Superfund lead agencies do not typically perform at contaminated sediment sites. However, biomonitoring of people who live near contaminated sediment sites is sometimes performed by other parties, such as local health authorities and academic scientists (Miller et al. 1991; Fitzgerald et al. 1996, 1999, 2004; MA DPH 1997; Korrick and Altshul 1998) and the Agency for Toxic Substances Disease Registry.

Typical indicators of human exposure and risk reduction include contaminant concentrations in the subset of environmental media to which people might be directly or indirectly exposed in places where exposures might occur. Those media include surface sediment and surface water in areas accessible to people and aquatic biota used for food. Where there is interaction between contaminated sediments and floodplain, the list may be expanded to include floodplain surface soil, terrestrial game species foraging in the floodplain, and agricultural products from the floodplain (see further discussion in sections below).

Use of Sediment Quality Guidelines

SQGs are numerical chemical concentrations intended to be either protective of biologic resources or predictive of adverse effects to those resources (Wenning and Ingersoll 2002). They are used to estimate the toxicity and risk from sediments. At a contaminated site, the SQGs can be used to establish contaminants of concern (COCs) from potentially long lists of contaminants of potential concern (COPCs), identify or rank problem reaches in a waterway, and classify hot spots (Long and Mac-Donald 1998).

There are two basic categories of SQGs, empirical and deterministic. Empirical approaches use statistical methods to compare sediment chemistry to effects datasets to predict the probability of or the presence and absence of adverse or toxic effects (Word et al. 2005). A variety of those approaches have been used to develop toxic effects levels, thresholds, or concentrations (used as SQGs) (MacDonald et al. 2000; Burton 2002; Wenning and Ingersoll 2002). Deterministic approaches typically use equilibrium partitioning theory (Adams et al. 1985; Di Toro et al. 1991, 1992) to relate toxic concentrations found in water-only exposures to sediment exposures for the same organism. Effects are predicted to occur when toxic concentrations found in water occur in the pore water of the sediment (Word et al. 2005); complexing agents (organic carbon for hydrophobic non-ionic contaminants [Di Toro et al. 1991] and AVS for cationic metals [Di Toro et al. 1990, 1992]) are the basis of the equilibrium calculations.

There has been a lot of controversy and discussion on the use and viability of SQGs including their false positive and negative rates, their applicability to mixtures of chemicals, their ability to establish cause and effect relationships, and whether results can be extrapolated across species or biologic communities (Burton 2002; Wenning and Ingersoll 2002). Sediment quality guidelines are only indirect measures of effects and do not clearly establish whether risk or adverse biologic impacts are actually occurring (Table 5-1) (NRC 2003).

A recent Pellston workshop summary, *Use of Sediment Quality Guidelines and Related Tools for Assessments of Contaminated Sediments* (Wenning et al. 2005), comprehensively reviews these approaches. A few conclusions reached from this workshop include that (1) although the scientific underpinnings of the different SQG approaches vary widely, none of the approaches appear to be intrinsically flawed; (2) chemically-based numeric SQGs can be effective for identifying concentration ranges where adverse biologic effects are unlikely, uncertain, and highly likely to occur, and; (3) in all cases, application of SQGs in a "toxic or nontoxic" context must be cognizant of the types and rates of errors associated with each type of SQG (Wenning and Ingersoll 2002).

EPA has supported the development of mechanistically based sediment quality guidelines (EPA 2000b, 2003a,b,c, 2005b) and the National Oceanographic and Atmospheric Administration (NOAA) has supported the development of empirical sediment guidelines (Long and Morgan 1990). It is expected that as the scientific issues continue to be resolved, they will see continued and greater use in toxicity evaluations, comparative risk analyses, and in remedial decision making.

MONITORING-PROGRAM DESIGN

Selection of the appropriate monitoring measures and design of a monitoring program depend on the development of clear hypotheses to be tested or questions to be answered that are directly linked to a detailed conceptual site model characterizing sources, pathways of exposure, and receptors that may be exposed during or after remediation. By the time a remedy is implemented, the understanding of site processes should be highly refined so that monitoring can be focused on the expected beneficial and adverse effects of remediation. These effects include releases to the water column or atmosphere, as monitored during dredging; residual sediment concentrations, as monitored by progress samples or post-dredging verification sampling; and reductions in exposures and risks, as observed through long-term monitoring.

Monitoring During Dredging

Dredging operations include material removal, transport, dewatering, final disposal, and onsite solids treatment, water treatment, and temporary storage. Therefore, monitoring during dredging operations may involve a variety of activities, including some that are not directly related to dredging operations or performance. An inherent difficulty is the need for rapid measurement techniques that can inform contingency actions and provide near-real-time feedback for executing corrective measures while the work is ongoing.

Monitoring programs implemented during dredging are often based on the requirements of Clean Water Act Rule 401 water quality certification, typically administered by state environmental agencies. The focus of any required monitoring for water quality certification is effects on water quality, based on comparison with state or federal water quality standards and criteria, taking upstream conditions into account. Surrogate or indicator parameters (such as turbidity or concentrations of a single chemical) are typically used to provide rapid information to the dredger and site manager and to develop a compliance history spanning the various phases of the project.

With available technologies, some contaminant release and transport is inevitable during dredging (EPA 2005a). Depending on the volatility of the contaminant, there may be release to the atmosphere as well as the water column. On the basis of project data presented in this report, contaminant release to the water column might not depend on observable resuspension of solids (see Box 4-2 for an example). Nevertheless, monitoring of turbidity can provide real-time quality-assurance information to the dredge operator and allow adjustments in the field to reduce resuspension. Air monitoring can also identify potential exposures and facilitate needed operational adjustments to protect nearby populations.

To quantify contaminant releases, however, upstream and downstream water-column contaminant fluxes should also be monitored. This can be accomplished relatively quickly using immunoassay test kits or traditional grab sampling with subsequent analyses. Passive sampling devices or caged fish can also be placed at the site to indicate exposure over extended periods. These techniques make it possible to quantify the unintended contaminant loading to the water column, and this helps to explain increases in downstream exposures that are observed between the baseline and long-term monitoring and to distinguish short-term effects due to dredging from continuing long-term releases attributable to uncontrolled sources.

Monitoring Human Health Effects During Dredging

During dredging and dredged-material handling, the surrounding community and remediation workers might experience higher exposures than before dredging. The increases could arise from chemical releases to the overlying surface water and ambient air, from uptake by biota consumed by people, and from creation of residual contamination in areas where people or edible biota come into contact with it. The surrounding community might also experience non-health-related effects, such as accidents, noise, and residential or commercial disruption, which are potential ancillary consequences of dredging.

An evaluation of net risk reduction should begin with sufficient datasets that permit comparison of exposure conditions before and during dredging operations. In some of the projects, increased exposure occurred during dredging in connection with the physical disruption of contaminated sediment. Monitoring during dredging should be designed so that data are sufficient to quantify changes in exposure resulting from the dredging operation, specifically related to resuspension of sediment, release of chemicals from sediment, and creation of residuals. Changes in net risk resulting from transport, storage, treatment, and disposal of dredged sediments should be quantified, and this may include collection of monitoring data during dredging and dredging-related operations.

Superfund remedial investigations often use concentrations of chemicals in environmental media—such as fish, sediment, surface water, and air—as surrogates for human exposure and do not study human subjects directly. Investigators need to monitor those media within the boundaries of the three-dimensional space in which people have direct or indirect contact. For example, people do not have direct contact with deep sediment. Unless that sediment becomes exposed in the future as a result of scouring, the dredging process itself, or some other process, sediment samples collected for evaluating direct contact should not exceed the depth that a swimmer or wader might encounter. For indirect exposure to stable sediments through the food chain, the relevant sediment sampling depth is limited to the biologically active zone. Sampling at greater depths is needed to assess potential exposures where sediments are unstable. If sediment contamination has reached the terrestrial environment through atmospheric release and deposition or sediment deposition on floodplain soils, parallel monitoring and risk analyses should be performed for terrestrial exposure media. Direct studies of human exposure could help to quantify human exposure and risk but would be more invasive and expensive and would not necessarily yield a good measure of exposure reduction, given the difficulty in defining the exposed population and segregating site-related exposures from other exposures to chemicals of concern.

Given that dredging remediation by definition involves an aquatic environment and that many of the most important sediment contaminants are bioaccumulative, the consumption of fish and other aquatic organisms often contributes most to human health risk. However, until dredging is completed and cleanup goals have been met, EPA and state agencies with fisheries jurisdiction usually restrict fish consumption to protect human health. When members of the surrounding community comply with those restrictions, exposures of concern during dredging are limited to other pathways, such as releases to the atmosphere and surface water. Box 5-3 highlights examples of attempts to evaluate human exposure during dredging.

Ideally, dredging operations occur over a relatively short period that requires evaluation of acute and possibly subchronic risk, but not chronic risk, from these exposure pathways. To address those risks, EPA can establish acute and subchronic guidelines for air or other media for comparison with monitoring results. For example, at the New Bedford Harbor Superfund site, EPA selected air concentrations that if exceeded during dredging would require a change in the dredging operation. Also at that site, EPA detected increased hydrogen sulfide concentrations in a dredged sediment handling facility and changed the operation to reduce concentrations to safe levels.[3] In such cases, monitoring results should be made available in a time frame that allows site managers to manage risks appropriately. The data should distinguish conditions upstream and downstream of dredging or upwind and downwind of dredging so that site managers can discern the effects of dredging relative to background exposure conditions including natural disturbances.

Monitoring Ecologic Effects During Dredging

Current practice often omits biologic monitoring during remedy implementation at sediment sites. Monitoring of bioaccumulation during dredging is typically not able to inform the project manager or operator in a timely fashion so that dredging protocols could be modified or additional protections implemented, owing to the length of time that most organisms take to respond to environmental exposures. Other challenges in using bioaccumulation and tissue concentration monitoring data are in relating chemical concentration to ecologic relevance or adverse biologic effects and in the uncertainty of the relationship between exposure (such as to site sediments or resuspended materials) and tissue concentrations in fish if they are able to move off site (these issues are described in greater detail in the next section).

---

[3]The presence of hydrogen sulfide is related to the anaerobic environment of the sediments and not to chemical contaminants at the site.

BOX 5-3 Monitoring of Conditions During Hot-Spot Dredging at the New Bedford Harbor Superfund Site for Effects on Human Exposure

The New Bedford Harbor Superfund site has been the subject of extensive efforts to understand the effects of harbor contamination on aquatic species and people living near the harbor. In addition, EPA developed a plan to monitor the effects of dredging a contaminated hot spot on water quality, air quality, and bioaccumulation by benthic invertebrates. The hot-spot dredging occurred in 1994-1995. EPA (1997) compared results of monitoring conducted before, during, and after hot-spot dredging and concluded that the dredging resulted in few if any adverse effects on the marine ecosystem. EPA identified some air-quality issues that were remedied with changes in operation or engineering controls.

Cullen et al. (1996) compared PCB concentrations in tomato samples collected downwind of the hot-spot dredging operation before and during dredging and concluded that the average PCB concentration during dredging was about 6 times higher than the average PCB concentration before dredging.

Choi et al. (2006) reported PCB concentrations in umbilical-cord blood samples among members of nearby communities that were collected before, during, and after hot-spot dredging. The authors reported that their results "support modest, transient increases in cord serum PCB levels during dredging, with significant declines in serum PCB levels observed after dredging, particularly for the more volatile PCBs and PCB-118." They attributed the "significant declines," in part, to the hot-spot dredging (see figure below).

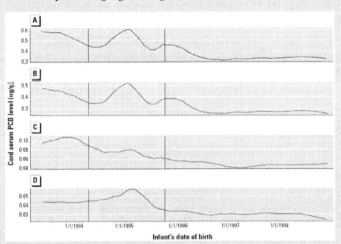

Figure shows covariate-adjusted smoothed plots of predicted $\sum$PCB (A), heavy PCB (B), light PCB (C), and PCB-118 (D) levels vs. infant's date of birth. Vertical lines denote the

(Continued on next page)

BOX 5-3 Continued

start and stop dates for dredging of contaminated New Bedford Harbor sediments. Plots are adjusted for child's sex, maternal age, birthplace, smoking during pregnancy, previous lactation, household income, and diet (consumption of organ meat, red meat, local dairy, and dark fish). Source: Choi et al. 2006.

EPA's summary of tier 1 sediment remediation sites with pre-monitoring and post-monitoring data (S. Ells, EPA, unpublished information, March 22, 2006) presented during the committee's first meeting indicates that the baseline average concentration of PCBs in surface sediment (no depth specified) at the hot spot was 25,000 mg/kg and lists a post-remedial average concentration the hot spot of 330 mg/kg. However, that information appears to be inconsistent with another EPA presentation (Nelson 2006) during which W. Nelson reported that the average concentration of PCBs in surface sediment (top 2 cm) of the upper harbor, of which the hot spot makes up about 5% of total area, did not change significantly as a result of hot-spot dredging. It is possible that the average was unchanged because the hot spot represents a small fraction of the upper harbor. However, if the hot spot represented the portion of the upper harbor with the highest PCB concentration, one would expect its removal to cause some decline in PCB concentration. Given that the PCB concentration apparently did not decline, it is not clear how hot-spot dredging might have led to reduced PCB concentrations in umbilical-cord serum after dredging.

These collective efforts show how exposure might change during the period of dredging, but it is premature to use them to judge effectiveness, because EPA's remediation is not yet complete. These studies illustrate the challenge of linking dredging in a large harbor with human exposure.

Nevertheless, one of the main risks to ecologic receptors posed by release and transport of contaminants during dredging is increased contaminant uptake and increased toxicity, and there are techniques for monitoring those effects, even if they are not in wide use. Subject to the timeliness limitations noted above for providing real-time feedback to dredging operations, studies can be designed to assess contaminant bioaccumulation and toxicity during dredging by using caged or sessile organisms or using passive sampling devices such as SPMDs, as discussed above (Chappie and Burton 2000; Adams et al. 2005; Crane et al. 2007). The utility of caged-fish studies has been demonstrated, for example, in the 1995 Grasse River non-time-critical removal action (BBL 1995). Mussels deployed in mesh bags have been used in the long-term monitoring program at the New Bedford Harbor Superfund site to monitor trends in PCB bioaccumulation and evaluate the impact of dredging operations (Bergen et al. 2005). To quantify the spatial distribution of resuspended materials, the organisms can be placed at various distances from the sources of contamination. Comparisons with pre-dredging, reference, or upstream conditions allow managers to determine whether uptake of contaminants increases during dredging operations.

Complex exposure dynamics cannot be mimicked in the laboratory. If standard test species are exposed in situ, exposures are more realistic. In situ testing with caged organisms has been shown to be an effective monitoring tool. Its primary advantages are the improved realism of exposure, the lack of sampling-induced artifacts, the ability to deploy and assess within days, and the ability to partition exposures of key environmental compartments and exposure time frames. (However, these techniques also have concerns regarding the modification of site conditions during exposure [Chappie and Burton 2000]). One can also link exposure with effects in that multiple end points can be assessed, such as tissue concentrations, growth, and reproductive status. Numerous studies have demonstrated the approach in studies of runoff, base flow, and sediments (e.g., Ireland et al. 1996; Tucker and Burton 1999; Chappie and Burton 2000; Greenberg et al. 2002; Burton et al. 2005; Crane et al. 2007). Studies of marine systems have primarily used mussels (Salazar and Salazar 1997; Bergen et al. 2005), and there has been less testing of amphipods (DeWitt et al. 1999). Freshwater studies have used a wide variety of organisms, such as fish, cladocerans, amphipods, midges, bivalves, mayflies, and oligochaetes (e.g., Chappie and Burton 2000). It is important to consider the likely response time when selecting test organisms; organisms that equilibrate with their environment more quickly would be more useful for evaluating releases during dredging. It is also advantageous, when possible, to use indigenous biota when conducting in situ caged testing. Standard toxicity test organisms (such as fathead minnows) may have very different biologic responses than indigenous populations that may have acclimated or adapted to toxics in the watershed, thus being less sensitive than the surrogate species. In that case, surrogate species may be useful for detecting adverse effects, but they will not be a good indicator of effects to indigenous species.

The use of natural resident populations collected during and immediately after dredging is an alternative to caging studies. For example, resident spottail shiners collected during the 2005 dredging operations at the Grasse River showed significant increases compared to sampling conducted during several years prior to dredging and the year following dredging (see Chapter 4, Figure 4-10 and associated text).

Post-dredging Verification Sampling

Verification sampling immediately after dredging allows site managers to determine whether cleanup levels or other short-term objectives have been met. The ability of the remedy to achieve short-term cleanup levels depends in part on how accurately the remedial investigation and additional pre-remedial sampling have characterized the extent and distribution of contamination and on whether the dredging design based on that characterization encompasses the bounds of contamination encountered by the dredger in the field. Dredging designs are based on interpolation of sediment core data, which are often sparse relative to the scale of sites. Even at major sites, the density of pre-design samples is typically less than one core per acre: for development of the Lower Fox River and Green Bay Operable Units 2-5 remedial design, the density was about one core per 1.6 acres (Shaw 2006). Depths of contamination in the wide expanses between core locations are therefore subject to uncertainty, and the dredging projects in this report provide evidence of that uncertainty in the form of sites where significant undisturbed residuals remained after dredging to design elevations. The probability of leaving consolidated sediments with elevated concentrations in place can be reduced by conducting more intensive pre-design sampling before dredging, by lowering the elevation of the dredge cut (that is, overdredging), or by verification sampling after dredging followed by redredging as needed (see examples in Box 5-4). Thus, there are tradeoffs between the volume of material removed and the intensities of pre-design and verification sampling. The greater the confidence in the methods used to develop the dredge prism, including sampling and interpolation, the less overdredging and verification sampling may be needed to ensure protection. Those tradeoffs have been considered explicitly in pre-design studies for Lower Fox River Operable Units 2-5 (see Box 5-5).

The dredging projects evaluated by the committee include numerous examples of sites where dredging generated substantial residual contamination. Verification sampling is needed to detect and quantify generated residuals. Where possible, the samples should be collected in the form of cores long enough to penetrate and capture sediment underlying the generated residual layer, rather than grab samples (Palermo 2006). When cores cannot be obtained, grab samples should be taken. One promising technology for obtaining grab samples, using a hydraulic sampling device, is described in Box 5-6. It is also important, during collection and analysis of cores, to capture the unconsolidated "fluff" that may be generated from dredging activity. With core samples, the thickness and texture of the generated residual layer can be observed and distinguished from underlying material to support planning of additional work that may be needed to minimize risk, such as backfilling with coarse-grained material.

BOX 5-4 Verification Sampling at Harbor Island, Washington

Dredging at Todd Shipyards (part of the Harbor Island Superfund site) relied on collection of shallow progress cores in each sediment management area (SMA) as dredging was completed. Results were compared with cleanup levels to determine whether additional dredging was needed. Dredging was sequenced in such a way that an SMA that had been remediated was not affected by dredging in adjacent SMAs. Final verification samples were collected once all SMAs had been dredged at a relatively low density because of demonstrated compliance with cleanup levels based on the progress cores.

The presence of extensive surface and subsurface debris in the areas to be dredged at Lockheed Shipyard (also part of the Harbor Island Superfund site) resulted in extensive residual contamination and undredged inventory at the end of the first dredging season. Shallow cores were collected throughout the dredged area after dredging to document the remaining contamination and distinguish between a light unconsolidated sediment layer and more consolidated material. The latter material either had sloughed from the edge of the dredge cut or could not be removed by the dredger because of debris that remained on the site. On the basis of this sampling, it was decided to place a thin layer of clean material over the dredged area until it could be redredged in the following season.

BOX 5-5 Delineating the Dredge Prism in the Fox River, Wisconsin

Sediment remediation areas and volumes were delineated for the remedial design of the lower Fox River, WI, between Operable Unit 2 and the mouth of Green Bay with an advanced interpolation method called full-indicator kriging. Full-indicator kriging provided a probability distribution of depth of contamination to the ROD cleanup level at each sediment location. For areas where dredging is the selected remedy, dredge-prism designs were developed at a range of significance levels (defined as the probability of exceeding the cleanup level at a given location) to inform risk management decisions. Those decisions involve balancing the risk of leaving contaminated sediment behind (a false-negative, or type 2, error) against the risk of unnecessarily dredging clean material (a false-positive, or type 1, error). Additional protection against false negatives will be provided in the remedy by post-dredging confirmation sampling. A significance level of 0.5 was chosen because it provided a reasonable Type 1 error and a low Type 2 error, reasonable accuracy, and the least bias in the dredge cut. This decision was made acknowledging the importance of minimizing Type 2 errors because remediation of clean sediments is cost that cannot be recovered. There was agreement that a robust verification sampling program would be developed and this would uncover significant Type 1 errors (that is, leaving behind sediments that should be remediated) which can subsequently be dealt with. In practice, more sediment will be removed than the selected significance level indicates because additional deepening of the dredge prism occurs during dredging-plan design and to account for contractor overdredging allowance (Anchor and Limno-Tech 2006a,b,c).

In verification sampling, it is important that the spatial scale of remedy evaluation be consistent with the site's remedial objectives. If the objectives require minimizing contaminant flux to the water column or minimizing sitewide exposures to widely ranging fish species, it is appropriate to compare area-weighted average concentrations with remedial goals. Protecting sensitive receptors that have more limited ranges would require verification that targets are achieved on a finer spatial scale. It should also be emphasized that although effectiveness of implementation, which is an intermediate goal, can be evaluated with verification sampling, the ultimate goal is risk reduction through achievement of remedial objectives, which is evaluated with long-term monitoring.

BOX 5-6 Verification Sampling of Dredging Residuals at the Head of Hylebos Site, Commencement Bay, Washington

Discrete sediment samples were collected on a daily basis immediately behind the operating dredge to provide immediate evaluation of the post-dredging residual layer for the Head of Hylebos project. Nearly 1,000 discrete samples of the residual layer were collected using a Marine Sampling Systems 0.3m² Power Grab (Power Grab) generating measurements of residual layer thickness and sediment chemistry (24-hour chemistry turn around times). This program provided immediate feedback on the nature of the residual layer generated during dredging, and allowed for ongoing adjustment of the dredging methods to further control the residual layer formation.

Unlike typical surface grab samplers, the Power Grab is a hydraulically actuated clamshell bucket that is capable of collecting 1-ft thick samples in many sediment types ranging from soft fine-grained sediment to more dense and compact silts and sands (not hardpan or glacial till), as well as through some debris. All of the sample contact surfaces on the Power Grab are stainless steel while the frame of the sampler is aluminum. The Power Grab features include (1) wide adjustable feet to control the depth of penetration to avoid over or under penetration of the sampler; (2) adjustable ballast (280-750 lb) to provide additional reaction weight for sampling in stiff material; (3) a semi-circular cutting profile to limit the disturbance of the sample; (4) and an enclosed bucket configuration to protect the sample from scour while being raised through the water column. These features allowed the sample team to consistently collect acceptable samples (without over- or under-penetration) in all sediment types found on the site.

Once the Power Grab sample was brought on board the sampling vessel, the overlying bucket covers were removed and overlying water decanted. The 0.3-m² sample footprint (roughly 1 3/4 ft by 1 3/4 ft) was sufficient to allow for subsampling to measure the thickness and record the characteristics of the residual layer, the characteristics of the underlying more compact native sediment, as well as the collection of sediment samples for chemical analysis. The Power Grab performed well throughout the two seasons of dredge confirmation sampling without any notable complications or problems.

Source: Dalton, Olmsted & Fuglevand 2006.

Long-Term Monitoring after Dredging

Long-term monitoring is used to judge whether a remedy is reducing risk at the expected rate and when remedial objectives have been achieved. Superfund remedies at sediment sites are typically subject to review at 5-year intervals when, following remediation, contamination exists that could limit potential uses of the site (EPA 2001b). At dredging sites this could occur for several reasons: residual contamination after the completion of the remedial action, the recontamination potential associated with the dynamic nature of the aquatic environment, the fact that some sources may be undetected and that controls of known sources are not always implemented concurrently with the remedy (particularly at the watershed level), and the additional time required by remedies to achieve objectives when they rely in part on natural recovery processes and must counter past bioaccumulation of contaminant in the food chain. Ideally, reviews compare recovery at each 5-year interval with an expected trend of exposure and risk reduction under the recommended remedy as developed in the feasibility study. Remedy modification and additional data collection as needed to fill in gaps in understanding, would be triggered by a significant deviation from the expected trend. Otherwise, monitoring continues until remedial objectives are achieved.

Although a rich set of data should already exist as a product of the remedial investigation, it is important that a complete baseline dataset be obtained before remedy implementation, observing the same pathways and exposures as planned for long-term monitoring, to support clear and definitive pre-remedial vs post-remedial comparisons and post-remedial trend estimates. The importance of establishing a baseline, especially to assess the effects of the remedy on fish, was emphasized by a previous National Research Council panel (NRC 2001). Because long-term monitoring may continue for decades and trend estimates will be based on comparisons with data collected in the early years of monitoring, it is important that sampling and analytic methods selected for baseline and long-term monitoring be consistent with the technologic state of the art. It is also vital that the baseline data be collected before the commencement of remediation and encompass trends of sufficient duration for the effects of the remedy to be distinguished from ambient trends leading up to its implementation. To facilitate comparisons of data over a long period (decades), it is useful to store duplicate biologic samples (fish tissues, human tissues, blood) from the analyses (for example in a deep freezer or liquid nitrogen). These samples can be analyzed in later years to facilitate comparisons of analytical data.

When a dredging remedy is implemented, surface sediment concentrations can be affected by a combination of sediment removal, backfilling with clean material, and natural recovery processes. At larger sites, where remediation may proceed over a period of years, there is value in determining the relative importance of each of those processes in reducing surface concentrations. If burial under clean watershed sediment transported by riverine processes strongly reinforces dilution of exposures, this may create opportunities to adapt the remedy to reduce cost without compromising effectiveness. For example, if burial by clean sediment is sufficiently rapid and uniform, it may be possible to achieve risk-reduction goals while tolerating higher generated residual concentrations or to reduce the thickness of post-dredging backfill (see Box 5-7). To be able to measure residuals, backfill, and long-term sedimentation as separate layers, baseline and long-term monitoring of surface sediment should include at least a subset of finely sectioned cores, which should be analyzed for geotechnical characteristics, including grain size, bulk density, and contaminant concentrations.

It is important to stress that backfilling and burial by natural sediment processes are not necessarily equivalent in protection even if they cover contamination to equal thicknesses. Sands, which are typically used as backfill material, may be more stable in the face of erosion during high-flow events than natural sediments but their lower sorptive capacity may provide less effective attenuation of contaminant exposure than burial by natural sediment, especially when pore water is the pathway of potential exposure.[4] Monitoring of surface concentrations should account for trends in bioavailability by normalizing contaminant concentrations to sorbent material, such as organic carbon, in addition to measuring trends in bulk surface-sediment contaminant concentrations.

Long-term Monitoring of Human Health Effects

Results of the baseline human health risk assessment indicate which exposure pathways and chemicals warrant action, and this knowledge is vital for making effective remedial decisions. As noted above, the consumption of fish and other aquatic organisms often contributes

---

[4] It also is possible to specify backfill material with organic carbon.

BOX 5-7 Dredging and Later Sedimentation at Manistique Harbor, Michigan

After dredging ended at Manistique Harbor, 3-7 ft of sediment were deposited from 1996 to 2005 (Weston 2005). With the deposition of the new surface sediment on post-dredging residuals, it was possible to meet revised dredging cleanup levels. The original cleanup level had been removal of all sediment containing PCBs at greater than 10 mg/kg anywhere in the sediment column, and it proved difficult to achieve. The cleanup level was revised to an average concentration of 10 mg/kg throughout the sediment column, with 95% removal of PCB mass also required (see Appendix C).

most to human health risk. Ideally, the population consuming aquatic biota would have been studied in the baseline human health risk assessment, including quantification of variability in fish consumption rates among members of the population to ensure that the most highly exposed members of the population are evaluated in the risk assessment. During monitoring, any important changes in consumption patterns should be accounted for in the monitoring plans. However, other pathways might be important, such as consumption of waterfowl, dermal contact with and ingestion of sediment, inhalation of volatilized contaminants, and ingestion of surface water during swimming or other recreational activities or through use as drinking water. If sediment contamination reaches the floodplain, people could be exposed through consumption of game species, wild edible plants, and agricultural products from the floodplain and through dermal contact with and ingestion of surface soil. Box 5-8 summarizes factors that one should consider in designing the aquatic sampling programs that are most often used to quantify human exposures.

EPA's goal at contaminated sediment sites is to protect human health, given that people could be exposed to sediment and other contaminated media over long periods. Consequently, cleanup levels, if established to protect human health, usually represent long-term average concentrations that people can be exposed to without expectation of harm over long periods. Therefore, they should not be treated as absolute exposure limits that are not to be exceeded at any time during long-term monitoring. Cleanup levels and monitoring programs should be

BOX 5-8 Collecting Aquatic Samples for Monitoring Human Exposure

Fish and Shellfish

- Sample the species commonly eaten by the local population and be sure to include species known to accumulate high concentrations of chemicals of concern.
- Catch the size range of fish harvested by the local population, being sure to include the larger fish usually harvested because larger (older) fish in a population are generally the most contaminated with chemicals that bioaccumulate (such as PCBs, dioxins, and methylmercury).
- Avoid sampling finfish species during their spawning period, because tissue concentrations of some chemicals (for example, such lipophilic chemicals as PCBs and dioxin but not methylmercury) may decrease during this time and because the spawning period is generally outside the legal harvest period.
- Match assumed or known consumption patterns to sampled species. Fish-creel data (from data gathered by surveying anglers) from state fisheries departments constitute one justifiable basis for estimating types and amounts of fish consumed from a given body of water. It is important to account for the fractions that various trophic levels contribute to a fish consumer's diet.
- Composite samples of fish parts consumed by the local population. People might eat skin-on fillets, skin-off fillets, or whole gutted fish (for example, in soups). Skin-off fillets will have the highest mercury concentrations, whereas whole-body fish samples will have the highest PCB and dioxin concentrations. PAHs do not tend to accumulate in finfish that metabolize them. Composites improve the chance of detecting chemicals and thus reduce the number of samples without detectable concentrations in the resulting dataset and the need to determine how they will be factored into arithmetic averaging.
- Use a probabilistic sampling design, randomly selecting sampling locations to address spatial variability and to ensure that sufficient samples are collected to distinguish the site from reference areas. This approach allows statistically valid inferences to be drawn on an area as a whole. Ideally, samples would be collected over a geographic area that represents the average exposure of those who eat fish from the body of water. If there are smaller areas where people are known to concentrate fishing, these areas should be intensively sampled.
- Collect both weight and length data to control for the potential influence of fish nutritional state on chemical concentration, such as by normalizing fish concentrations to a standard body condition.

(Continued on next page)

BOX 5-8 Continued

Sediment

• Collect from accessible locations where people are likely to fish, swim, or engage in other activities (sediment samples in deep water, for example, may be relevant to the food-chain exposure pathway but not the direct-contact pathway).

• To evaluate direct-contact exposures, collect sediment at depths that correspond to the depth to which a swimmer or wader might sink.

• To evaluate indirect food-chain exposures, collect sediment from the biologically active zone.

Surface Water

• Collect from accessible locations where people are likely to fish, swim, or engage in other activities.

• Collect from areas used as a drinking water source.

• Measure total chemical concentrations if people ingest the water. Dissolved-phase concentrations are more useful for some evaluations of dermal exposure.

Source: Adapted from EPA 2000c.

defined in the context of areas over which people average their exposure. For example, if the pathway of concern is direct contact with sediment, concentrations in sediment that are routinely beneath 10 ft of water are of less concern than concentrations in shallow, accessible waters at the shoreline. Cleanup levels should be compared with uncertainty bounds on average exposures, such as the 95% upper confidence limits of the mean concentration in each human exposure area, rather than the maximum concentration detected in each exposure area.[5]

The conceptual site model and feasibility study results should be used to set expectations for the rate of risk reduction. To ensure that unacceptable risks do not occur, site managers can track concentrations

---

[5]One caveat to the safety of long term averages and the relative unimportance of short term exceedances is that exposure can occur during a vulnerable period, such as fetal development or infancy.

monitored over time to estimate expected cumulative cancer risk and noncancer hazards. If monitoring during or after dredging indicates that cleanup levels will not be met in the long run, site managers can use adaptive management to change this trend.

Long-term Monitoring of Ecologic Effects

A primary goal of long-term monitoring is to test the hypothesis that dredging has reduced injury to ecologic resources. Many of the techniques used to assess potential short-term adverse effects of dredging on ecologic resources, described above, are also appropriate for assessing long-term effectiveness of sediment removal. For example, benthic toxicity testing has been successfully used as part of long-term monitoring at the former Ketchikan Pulp Company site in Alaska. In Waukegan Harbor, laboratory sediment toxicity assays conducted four years after dredging showed reduced toxicity compared to pre-dredging assessments, but nevertheless, toxicity still persisted (Ingersoll et al. 1996; EPA 1999; Kemble et al. 2000). Followup studies in the Black River showed that surficial sediments had reduced toxicity but PAHs were still causing toxicity in caged organisms (Burton and Rowland 1998). Contaminant uptake and toxicity can also be quantified by using in situ approaches with caged organisms or using passive sampling devices such as SPMDs, as discussed above. In addition to placing cages or SPMDs at different locations to quantify spatial distribution of contaminant concentrations or effects, observations can take place over time to determine temporal changes in bioavailability and sediment toxicity. SPMDs were used in Manistique Harbor 4 years after dredging operations ceased. The SPMDs accumulated PCBs to detectable levels while PCBs were not detected in caged fish or surface water samples at the site (Weston 2005).

Long-term monitoring of resident populations of fish and invertebrates can also reveal changes in contaminant concentrations and ecologic effects resulting from removal of contaminated sediment and natural processes. For example, the incidences of lesions and tumors in brown bullheads in the Black River showed initial increases and then marked reductions following dredging at the site (Baumann 2000) compared with pre-dredging conditions. Tissue data from fish whose habitat is limited to the remediation site are valuable indicators because they integrate exposures over the remediation area. Several of the dredging projects (such as Waukegan Harbor, Grasse River, Black River, the Puget Sound Naval Shipyard, and GM Massena) that were evaluated by the committee monitored fish tissue concentrations of contaminants to evaluate the effectiveness of sediment removal. Distinguishing the effects of remediation from background trends on the basis of fish tissue data often proves problematic for the decision-making process because of the scarcity and variability of fish tissue data. In addition, the difficulty in quantifying the movements of fish in and out of a project area can make linkages between exposure and effects problematic. It may be impossible to determine how much time a fish has been exposed to study-site sediments compared with offsite sediments (which may also be contaminated). Bioaccumulation modeling approaches (e.g., Linkov et al. 2002) that include spatial and temporal characteristics of exposures based on a knowledge of the organism's life history patterns can be useful in addressing that issue. Furthermore, pre-dredging data are often limited to a short time and very few fish (such as at Waukegan Harbor), so it might be impossible to make statistically valid comparisons of trends (as discussed in the next section). In these situations, caged-fish studies can maximize exposure to test-site sediments and thereby reduce uncertainty. Fish may be the receptors of primary interest for both ecosystem and human health risk, but monitoring them supports effective decision-making only if sufficient samples are collected and their patterns of exposure are known.

For determinations of ecosystem and human health risk, it is sometimes more effective to monitor tissue concentrations in benthic invertebrate organisms that reside at the test site (Adams et al. 2005; Burton et al. 2005; Solomon et al. 1997) because these organisms tend to be sessile or relatively immobile (for example, mussels). The organisms are exposed to the contaminated sediments through direct contact and are a food source for fish, birds, and mammals; therefore, food-web transfer and risk can be (and have been) modeled. The uncertainty of exposure is largely removed, and organisms are easier and less expensive to collect than fish and provide a convenient surrogate, as long as any assumed bioaccumulation link can be verified with site data. In addition, passive sampling devices that are biomimetic have recently been successfully used (see above discussion). The adsorption of organic and metal contaminants on these devices has been shown to be similar to that of tissue concentrations in indigenous organisms, so they can be used as a surrogate for fish (Arthur and Pawliszyn 1990; Huckins et al. 1990; Zhang et al. 1998; Wells and Lanno 2001; Lanno et al. 2004, 2005). Bioaccumulation and toxicity studies can also be conducted in the laboratory with sediment and water collected from field sites after dredging operations. However, it is critical that laboratory studies consider abiotic factors that may influence contaminant bioavailability and degradation—such as ultraviolet light, suspended solids and colloids, and organic carbon— and the effect that removing sediments from the environment will have on bioavailability.

Monitoring the structure and composition of benthic macroinvertebrate communities is a common approach to the assessment of effects of sediment contaminants and can be used when the benthic community is an important component of the conceptual site model of increased site risks. Such community characteristics as total abundance, species diversity, richness, and abundance of sensitive species can be compared with pre-dredging data and, when possible, with nearby reference sites. Again, it is critical that similar methods be used to collect and process pre-dredging and post-dredging samples. One of the greatest challenges associated with long-term monitoring of benthic communities is to separate effects of dredging from changes due to other environmental factors. The condition of benthic communities is generally expected to improve after the removal of contaminated sediment, as may be predicted by the conceptual site model. The failure of benthic communities to recover after dredging could be a result of residual sediment contaminants, lack of colonizing organisms, conversion to an inhospitable or unsuitable habitat, or the presence of other stressors (Kelaher et al. 2003). Monitoring approaches using in situ cages that contain natural benthic communities offer an opportunity to demonstrate causal relationships between stressors and ecologically relevant responses. Demonstrating changes in the tolerance of populations or communities may also provide evidence of effectiveness of dredging in situations where traditional community metrics (such as abundance and species richness) do not show recovery. For example, increased tolerance to metals is often observed at metal-contaminated sites (Weis and Weis 1989; Clements 1999), so the loss of tolerance in a population or community after dredging is evidence that remediation was successful (Levinton et al. 2003). These experiments are a practical alternative to single-species toxicity tests and address the statistical problems associated with field biomonitoring studies (Clark and Clements 2006).

The rate of recovery of benthic communities after dredging will be determined by both biotic and abiotic factors (Yount and Neimi 1990). The rate of recovery will be influenced not only by the adverse effects of large-scale substrate disturbance and the presence of residual contaminants, but also by proximity to reference areas and the availability of colonizing individuals. Ecosystems that have a direct connection to clean reference sites will probably recover faster than closed systems that have relatively little exchange. For example, the relatively fast recovery of stream ecosystems after remediation has been attributed to rapid colonization by organisms from upstream reference sites (Clements and Newman 2002).

DATA SUFFICIENCY AND STATISTICAL DESIGN

Monitoring datasets should be rich enough to support testing of the hypothesis that dredging is effective in meeting its remedial goals and objectives. That requires that sampling targets the important exposure pathways and be designed to capture temporal and spatial variability and that sample sizes be sufficient for robust hypothesis-testing and statistical modeling of dredging effectiveness goals.

Standard statistical tests are often formulated as a null hypothesis representing no effect or no change vs an alternative hypothesis representing an effect or change. When evaluating dredging effectiveness on the basis of pre- and post-dredging data, the null hypothesis represents no change due to dredging; the alternative hypothesis is that there was a change in environmental conditions because of dredging. Established formulas exist (EPA 2000d) for sample size determinations based on that traditional approach. With the required estimate of outcome variability (as can be obtained, for example, in a pilot study) and specification of the minimal effect size that should be detected, sample size determinations are based on optimizing the two types of statistical errors that can result. The probability of type I errors (incorrectly claiming dredging to be effective when it was not) is fixed to be small, as is commonly done by setting the probability at 0.05. The probability of type II errors (failing to claim that dredging was effective when it really was) is minimized, maximizing statistical power (Mason et al. 1989). The approach is thus conservative; the burden of proof is on the monitoring data to provide enough data points to support dredging effectiveness with a high degree of confidence. One-sided alternative hypotheses can also be considered to test whether an effect or change was in a specific direction, such as a significant reduction in site conditions, and can consider varying minimal effect sizes, such as a reduction in site conditions of at least 90% from pre-dredging or related background values. The latter approach can be compared with hypothesis-testing techniques based on the bioequivalence paradigm (McDonald et al. 2003).

Sample size determination is crucial, but other components of statistical experimental design should not be overlooked in developing monitoring plans to evaluate dredging effectiveness. A clear scientific definition of dredging effectiveness is needed so that appropriate statistical hypotheses can be formulated. Hypotheses to be tested should be based on and fully informed by the conceptual site model of exposures and risks, as developed in the remedial investigation and baseline risk assessment. Outcome variables need to be established, and their spatial and temporal support (where, when, and how much) should be determined; all this should be consistent with and inform the statistical hypotheses. Careful determination and measurement of potential sources of variation that may affect outcome variables are also important. Two sources of variation that deserve further focus are temporal and spatial variation in dredging effectiveness and their influence on monitoring and followup statistical analysis.

It has been well established in this chapter and in the dredging projects reviewed in the previous chapter that characterizing environmental conditions with monitoring before and after dredging is an important design consideration for evaluating dredging effectiveness. Less established are guidelines for determining when temporal characterizations should be assessed and whether assessment should follow a crosssectional approach of one time before and one after dredging or be longitudinal and use multiple monitoring times before and after dredging. With just two time monitoring points before and after dredging (with multiple samples taken at each of these time points), one can determine whether a significant increase or decrease occurred between the two time points. When dredging (or another remedial action) takes place between these points, it is often assumed that the change results from the remediation, however, it is essential to consider trends that would occur regardless of dredging (for example, natural decreasing or increasing trends in contaminant concentrations in sediment or fish). The value of monitoring at several time points before and after dredging is that any trends not due to dredging can be determined and the effect of dredging more clearly established. Examples of fish tissue analyses that would have benefited from more complete time trend data are presented in Chapter 4 Boxes 4-1 and 4-6 on the Grasse River and Waukegan Harbor, respectively.

Spatial or geographic variability can be an important component of overall variability to consider in designing monitoring plans to evaluate dredging effectiveness. There could be several reasons for spatial variation in dredging effectiveness at a site. There could be naturally occurring variations in environmental factors, such as water flow, wind patterns, and sediment texture. There could be spatial variations in site conditions that affect the ability to dredge or dredge effectively, such as the presence of bedrock, harbor infrastructure, or debris. It is not only important to collect location information with monitoring data but to statistically inform the monitoring plan to determine appropriate locations of monitoring samples. For example, in analysis of the Grasse River project (see Chapter 4), locations of sediment samples were too far apart to identify spatial variation in surficial PCBs. Analyses of the Lavaca Bay project (see Chapter 4) suggested that subarea variations in surface mercury were of interest, but samples were too small to test this hypothesis statistically while also considering temporal variation after dredge passes. The subfield of statistics known as geostatistics (for textbook treatments, see Cressie 1991; Goovaerts 1997; Diggle and Ribeiro 2007) deals with the design and analysis of spatially referenced data that commonly arise in monitoring of dredging applications and should inform monitoring plans and data analysis.

Even when spatial variation in dredging effectiveness is not of primary interest, the data collected through monitoring may very well exhibit spatial dependence, that is, measurements of samples taken closer together are more similar than those of samples taken farther apart. Overlooking that property can result in hypothesis tests and statistical-model inference with biased levels of significance (Cressie 1991). Obtaining and including sample coordinates in monitoring databases will allow followup statistical analyses to include possible spatial dependence and make the appropriate adjustments when necessary.

In the dredging projects reviewed, the committee found that the quantity and quality of available and accessible monitoring data varied considerably. Followup statistical analyses of monitoring data often were nonexistent or consisted of simple summaries and graphs lacking any formal notion of statistical uncertainty; this created a critical gap between the large expenditures devoted to monitoring and the ability to provide scientifically defensible claims of dredging effectiveness based on monitoring data. It is imperative that rigorous statistical analysis of monitoring data be performed so that assessments of dredging effectiveness reflect the inherent uncertainties involved.

APPROACHES TO IMPROVING MONITORING

The dredging projects reviewed by the committee revealed limitations in the ability to make real-time adjustments in dredging operations to minimize contaminant releases; to connect remedial actions with their effects in space and time on exposure pathways, receptors, and ecosystems; to base monitoring on adequate conceptual site models of chemical fate and transport and of human health and ecologic risk; and to understand the roles of multiple processes in determining effectiveness of dredging. This section proposes approaches with promise to overcome those limitations. Each will require method development and evaluation before becoming part of the standard monitoring tool kit, and some may be appropriate only in particular cases. The methods can contribute to a weight-of-evidence basis of decision-making (Wenning et al. 2005) to reduce uncertainty in evaluation of risk reduction at specific sites. The topic of innovative monitoring methods is also reviewed by Viollier et al. (2003) and Apitz et al. (2005).

Approaches with potential to improve site investigation and operational and post-remedial monitoring include the following:

- Measure sediment, pore water, and surface-water concentrations rapidly and accurately.
- Monitor real-time contaminant releases during dredging.
- Measure the bioavailable fraction of contaminants in the field.

- Closely link exposure data (that is, chemical data) with biologic effect.
- Understand and model biologic uptake.
- Understand and model ecosystem response and recovery.
- Understand and model reduction in human exposure.
- Adequately and quickly identify generation, production, transport, and deposition of sediment residuals.
- Understand and model processes responsible for recovery after dredging.
- Quantitatively account for uncertainty in predictions of risk reduction and in later monitoring.

Measure Sediment Pore Water Concentrations Rapidly and Accurately

A growing body of evidence suggests that sediment pore water concentrations are strong indicators of the effects of sediment-contaminant concentrations on benthic organisms (Adams et al. 1985; Di Toro et al. 1991; Jager et al. 2000; Kraaij et al. 2003; Wenning et al. 2005; Lu et al. 2006). Sediment pore water concentration is directly related to the amount of bioavailable contaminant and uptake by benthic organisms (McLeod et al. 2007). Current methods to measure sediment pore water involve the equilibration of sediment samples in the laboratory and extraction of equilibrated water or the use of biomimetic assays, as discussed above. Rapid techniques for measuring sediment pore water would provide more useful and timely information on the status of recovery and resulting reduction in risk to humans and ecosystems. Method development and pilot testing are needed to determine how reliably the available techniques can be adapted for laboratory and field conditions.

Monitor Real-Time Contaminant Releases in the Field During Dredging

Cost-effective methods are needed for real-time monitoring of contaminant releases during dredging. At present, turbidity commonly is used as a surrogate for the release of persistent organic contaminants (such as PCBs) because the measurement is robust and quick and may be automated. It is often assumed that turbidity release is proportional to contaminant release. However, that may not be the case, as was shown in several of the Chapter 4 dredging project evaluations, because contaminant fractionation between the aqueous phase and sediment particles can result in releases of aqueous-phase (or colloid-associated) chemical contaminants, depending on the size and chemistry of the solid phase. Furthermore, if dredging exposed nonaqueous phases, such as liquid tar or hydraulic oil, contaminant release from such phases to overlying water would not be related to the release of solids. In principle, some of the latest methods to measure contaminants in sediment pore water—such as ELISA, PEDs, and SPME—could be applied to monitor the release of contaminants to the aqueous phase if their detection limits prove adequate. The newer methods in conjunction with turbidity, to the extent that they are correlated, may facilitate more reliable and faster contaminant monitoring during dredging.

Measure the Bioavailable Fraction of Contaminants in the Field

A variety of potential methods are candidates for measuring the bioavailable fraction of organic and inorganic contaminants in sediments. This information, for both baseline and post-remedial sampling, would complement chemical data to provide a more complete picture of changes in exposure due to the remedy. The National Research Council report on the bioavailabilty of contaminants in soils and sediments has a long and detailed chapter devoted to this topic (NRC 2003). However, most of the methods are not compound-specific or require detailed instrumental methods of analysis. Field methods that are compound-specific are desired. One promising approach is immunoassay techniques for assessment of the bioavailable contaminant concentration. Contaminant-specific ELISAs (Johnson and Van Emon 1996) may be rapid, useful tools for measuring available contaminants. The immunoassay uses the selectivity and sensitivity of antibody recognition coupled to an enzymatic reaction to rapidly determine chemical (such as PCB) concentrations in a variety of media, including wet sediment extracts and pore water. Ideally, the whole procedure, from extraction to colorimetric detection, could be carried out in the field with a test kit and portable equipment [Ta 2001].

Understand and Model Biologic Uptake

Site models that are used to support site investigations, remedy selection, and remedial monitoring should include a model of contaminant uptake by the affected biota. Biodynamic models describe the uptake of contaminants as a mass balance of uptake from water; uptake from food particles, including sediment; and loss rates. Such models would help to explain the relationship between level of sediment cleanup and concentration of contaminants in organisms. The typical bioenergetics-based toxicokinetic model (for example see Norstrom et al. 1976) assumes that uptake by each route is independent and additive. The model has been used to determine the uptake of contaminants by different routes with experimentally determined model parameter values (Boese et al. 1990; Weston et al. 2000; Lu et al. 2004). Luoma and Rainbow (2005) recently proposed biodynamics as a unifying concept in metal bioaccumulation, and similar formulations have been used in PCB food-web models (e.g., Connolly and Thomann 1992). It is proposed that a biodynamic model that integrates sediment, water, and organism data from field projects with the rapid assessment techniques described above be used to predict contaminant concentrations in several species of interest. For example, McLeod et al. (2007) showed that this biodynamic model successfully predicted PCB body burdens in the clam *Macoma balthica* exposed to untreated and activated-carbon-amended Hunters Point sediment in laboratory experiments (see Box 5-9).

Understand and Model Ecosystem Response and Recovery

We lack rigorous modeling approaches to predict the ecologic characteristics of recovery after sediment cleanup. It is possible that an explanatory approach to ecologic recovery could build on the biodynamic modeling described above. If so, it could be incorporated into the conceptual site model and used to support the site investigation, remedy selection, and post-remedial monitoring. That belief is founded on the

---

BOX 5-9 Biodynamic Modeling to Predict Organism
PCB Concentrations

If an organism is considered a single compartment for contaminant uptake, the following biodynamic equation describes its accumulation of a toxic
contaminant (McLeod et al. 2007):

$$\frac{dC_{organism}}{dt} = FR \cdot AE_{aq} \cdot C_{aq} + IR \cdot AE_{sed} \cdot C_{sed} - k_e \cdot C_{organism}$$

where, $C_{organism}$ is the contaminant concentration in soft tissue (µg/g dry), FR is
the water filtration rate (L of water per g dry per day), AE_{aq} is the contaminant
absorption efficiency from water, C_{aq} is the aqueous contaminant concentration
(µg/L), IR is the sediment-particle ingestion rate (g of sediment per g dry per
day), AE_{sed} is the contaminant absorption efficiency from sediment, C_{sed} is the
sediment contaminant concentration (µg/g dry), and k_e is the proportional rate
constant of loss (per day). Model parameters include organism and filtration and
ingestion rates estimated from the literature. Sediment and aqueous contaminant
concentrations would be measured in situ, and laboratory experiments would
determine absorption-efficiency values and loss rates for the model organisms.
The advantage of this approach is that once the organism parameters values are
obtained, the conceptual model is transferable to other locations.

---

fact that contaminants in sediments simplify community structure by
eliminating some species but not others. Therefore, recovery should involve return of the contaminant-sensitive species to the community. In
addition, benthic communities in estuaries are dynamic in space and
time (Nichols and Thompson, 1985), so traditional ecologic observations
should be frequent and detailed to resolve community recovery. Biodynamic modeling based on functional ecology may allow prediction of the
species most sensitive to a contaminant, and this predictive capability
will help biologists to identify which species from the available recruitment pool are likely to recolonize a site when the contaminant is removed or bioavailability is reduced. Recolonization of the contaminant-
sensitive species in a recovering habitat reflects the success of remediation. A hypothesis to be tested is that recolonization predictions can be
built from basic information on taxon-specific functional ecology and

biodynamics and contaminant metabolism, combined with data on species availability for community recruitment.

Understand and Model Reduction in Human Exposure

Monitoring programs should include measurement of surface sediment, surface water, edible aquatic species, and other environmental media found in the baseline risk assessment to present unacceptable human health risks through either direct or indirect exposure. Monitoring determines whether exposure concentrations have declined as predicted. In addition, some systematic studies of the U.S. population, such as the National Health and Nutrition Examination Survey (NHANES), include biomonitoring data on some of the contaminants commonly detected at Superfund megasites. At the New Bedford Harbor Superfund site, members of the surrounding community have been studied, including collection of umbilical-cord serum and breast-milk samples, as part of an epidemiologic study of PCB effects on young children. Such human biomonitoring studies can be expensive and invasive and are not entirely without risk to those being monitored. Therefore, the committee does not recommend implementing human biomonitoring sampling for all dredging projects. However, if relevant human biomonitoring data exist, they can be reviewed for evidence that dredging resulted in reduced human exposure and risk. Noninvasive biomonitoring might also improve future assessments of human exposure. For example, Fitzgerald et al. (2005) reported a significant correlation between a noninvasive test of enzyme activity related to PCBs and serum concentrations of PCBs in members of a Mohawk tribe living near the General Motors-Central Foundry Division Superfund site along the St. Lawrence River. Serum PCB concentrations in this population had previously been correlated with consumption of fish from the river (Fitzgerald et al. 1996, 1999, 2004).

Adequately and Quickly Identify Generation, Production, Transport, and Deposition of Sediment Residuals

The purpose of sediment verification sampling is to ascertain whether additional dredging passes, backfilling, or other remedial followup is needed to meet risk-based cleanup levels. Downtime for dredging equipment and operators is expensive, but verification sampling and laboratory analysis can be slow and laborious and require dredgers to move on to other locations and return when results are available and have been reviewed. Methods include grab sampling, coring, and visual inspection by diver or with an underwater camera.

Operator response to verification sampling is limited by best achievable laboratory- turnaround times and would be improved by development of more reliable field methods of analysis and greater use of mobile laboratories. In combination with the most rapid methods of analysis, the development of correlations between target chemical concentrations and sediment physical properties, as may be reflected in sediment layering and other geomorphologic features, has the potential to streamline sampling and analysis and to provide cost efficiencies and much more rapid feedback to operators (Dow 2006).

Understand and Model Processes Responsible for Recovery After Dredging

As discussed above, a dredging remedy can affect surface sediment concentrations through a combination of sediment removal, backfilling, and enhancement of natural recovery processes by creating areas of preferential settling and deposition. Backfilling in particular was a component of the remedy at many of the sites considered in Chapter 4, but its risk-reduction efficacy is uncertain and probably depends on the nature and thickness of the backfill material. Although backfilling provides a separation layer between the water column and the contaminant, the effective attenuation of exposure by backfill material may be minimal if it is in a thin layer and has low adsorptive capacity.

To understand the long-term effects of dredging remedies on risk, those issues should be evaluated as part of the monitoring program. Because undredged residuals, generated residuals, and backfill material would be expected to differ in grain size and bulk density, these layers should be delineated, after dredging, through physical and chemical analysis of finely segmented cores. A time series of similar followup coring data, as part of the long-term monitoring program, would suffice to distinguish the effects of post-dredging burial from those of backfilling.

To estimate the combined effects of backfilling and burial on bioavailability, organic-chemical concentrations in surface layers should be organic-carbon normalized, and acid-volatile sulfide analyses of metals should be conducted. Emerging field methods of pore water and bioavailability analysis should also be applied as they become more reliable and widely available.

Account for Uncertainty Quantitatively in Predictions of Risk Reduction and Later Monitoring

All risk assessments have inherent uncertainty in fate and transport modeling and quantification of exposure and toxicity. The goal of monitoring should be to measure a given level of net risk reduction with a reasonable degree of confidence. By acknowledging uncertainty, one is better equipped to design an effective monitoring program and to answer questions from affected communities. For example, quantification of uncertainty may enable site managers to inform community members that the vapor-phase concentration of a chemical that will be released during dredging operations and reach the nearest neighborhood, on the basis of the best available modeling, is well below levels of concern at a specific high level of confidence.

CONCLUSIONS

The committee draws the following conclusions concerning monitoring of dredging effectiveness in reducing risk:

- Monitoring is the only way to evaluate the success of a remedy in reducing risk and is therefore an essential part of the remedy.
- Trends that occur at these sites are subject to biologic, chemical, and physical processes that often operate on long time scales. The trends and processes may be best described and understood with long-term modeling and monitoring in pre-remedial and post-remedial time frames.
- In the absence of sufficient baseline data, it is impossible to evaluate effectiveness. Where pre-dredging conditions are not static, a

pre-remedial time-trend analysis is needed to judge remedial effectiveness.

- In most cases reviewed by the committee, monitoring has not been adequately designed or implemented. Specifically,
    - o The design of the monitoring has often not been linked sufficiently to the conceptual site model.
    - o Tools developed for the remedial investigation, including numerical models and baseline risk assessments, are often neglected in formulating monitoring plans.
    - o Baseline datasets have not always been consistent with long-term monitoring data.
    - o Contaminant exposure and effects have not always been adequately linked in time and space.
    - o In many cases, the quality and quantity of monitoring have been insufficient to support rigorous statistical analyses.
- Some of the currently used monitoring techniques have proved useful in determination of short-term and long-term effects of remediation. These include:
    - o Monitoring during dredging, such as measurement of mass flux through upstream and downstream chemical monitoring and biologic monitoring, including caged-fish studies.
    - o Long-term monitoring of fish tissue, where appropriate, and other pathways that contribute substantially to human health risks.
    - o Long-term monitoring of affected benthic communities, including tissue concentrations and health of benthic communities.
    - o Laboratory toxicity testing using benthic organisms in sediment to monitor long-term changes following dredging.
- If fish are exposed to offsite conditions, there is uncertainty as to their exposure and the relationship to risk. In those cases, benthic organisms may be better indicators of exposure, provided that their use is consistent with the conceptual site model of exposure pathways. If biologic testing is not possible, passive sampling biomimetic devices provide indications of contaminant exposures.

RECOMMENDATIONS

The committee offers the following recommendations for improving monitoring of dredging effectiveness:

- EPA should ensure that monitoring is conducted at all contaminated sediment megasites to evaluate remedy effectiveness. That will require a commitment of resources commensurate with the scale and complexity of the site.
- Monitoring plans should focus on elements required to judge effectiveness and inform management decisions for the site. Care should be taken to select the correct indicators of ecologic or human risk carefully. All aspects of monitoring—including planning, evaluation, and adaptive management based on monitoring findings—should be closely linked to the to conceptual site model so that the hypotheses and assumptions that led to the selected remedy can be tested and refined.
- The breadth and richness of monitoring datasets should be sufficient to support the testing and full evaluation of effectiveness goals. Statistical expertise should be included to inform well-designed monitoring programs, guide database development, and perform rigorous statistical analysis of monitoring data aimed at evaluating effectiveness.
- EPA should ensure that monitoring information on all Superfund megasites is systematically collected, organized, analyzed to assess the effectiveness of remediation, and made available to the public in such a form that effectiveness evaluations can be independently verified.
- Numerical models that are used in the remedial investigation and feasibility study to design the remedy should be revisited during the remediation phase to help in the evaluation of the effectiveness of remediation.
- If possible where combination remedies have been used, the relative contributions of dredging, capping and backfilling, and natural recovery should be measured through sediment monitoring, and the results of monitoring should be used to adapt and optimize remedies.
- Remediation decision makers should examine the expected net risk reduction associated with each remedial alternative before selecting a remedy that will be implemented and link the monitoring program to the assessment of net risk reduction for the selected remedy.

• Pre-remediation baseline monitoring methods and strategies should be developed to allow statistically valid comparisons with future monitoring datasets that rely on time-series data. The ultimate goal is to assemble a consistent long-term dataset for conducting evaluations. During preliminary and final remedy design, monitoring should be initiated to help to establish a time trend integrating earlier characterization data as technically appropriate.

• Monitoring of the benthic community, as surrogate species reflecting food-web transfer, can provide valuable information about ecosystem health and integrate short-term and long-term exposures and multiple life stages. When consistent with the conceptual site model of exposure and risk development of site-specific approaches is recommended.

• Faster and less expensive monitoring methods that are deployable in the field are needed to better inform dredging operations in real time and to improve predictive capability.

REFERENCES

Adams, R.G. 2003. Polyethylene Devices and the Effects of Sediment Resuspension on the Cycling of PAHs and PCBs in the Lower Hudson Estuary. Ph.D. Thesis, Massachusetts Institute of Technology, Cambridge, MA.

Adams, W.J., R.A. Kimerle, and R.G. Mosher. 1985. Aquatic safety assessment of chemicals sorbed to sediments. Pp. 429-453 in Aquatic Toxicology and Hazard Assessment: Seventh Symposium, R.D. Cardwell, R. Purdy, and R.C. Bahner, eds. STP 854. Philadelphia, PA: American Society of Testing Materials.

Adams, W.J., A.S. Green, W. Ahlf, S.S. Brown, G.A. Burton, B. Chadwick, M. Crane, R. Gouguet R, K.T. Ho, C. Hogstrand, T.B. Reynoldson, A.H. Ringwood, J.D. Savitz, and P.K. Sibly. 2005. Using sediment assessment tools and a weight of evidence approach. Pp. 163-226 in Use of Sediment Quality Guidelines and Related Tools for the Assessment of Contaminated Sediments, R.J. Wenning, G.E. Batley, C.G. Ingersoll, and D.W. Moore, eds. Pensacola, FL: SETAC Press.

Anchor and LimnoTech. 2006a. Evaluation of Geostatistical Methods for Delineating Remediation Boundaries in Operable Unit 3, Lower Fox River Remedial Design, OU 2-5. February 28, 2006.

Anchor and LimnoTech. 2006b. Evaluation of Geostatistical Methods for Delineating Remediation Boundaries in Operable Unit 4, Lower Fox River Remedial Design, OU 2-5. February 28, 2006.

Anchor and LimnoTech. 2006c. Supplemental Information on Geostatistical Performance Metrics, Operable Units 3 and 4, Lower Fox River Remedial Design, OU 2-5. February 28, 2006.

Apitz, S.E., J.W. Davis, K. Finkelstein, D.W. Hohreiter, R. Hoke, R.H. Jensen, J. Jersak, V.J. Kirtay, E.E. Mack, V.S. Magar, D. Moore, D. Reible, and R.G. Stahl, Jr. 2005. Assessing and managing contaminated sediments: Part 2. Evaluating risk and monitoring sediment remedy effectiveness. Integr. Environ. Assess. Manage. 1(1):e1-e14.

Arthur, C.L., and J. Pawliszyn. 1990. Solid phase microextraction with thermal desorption using fused silica optical fibers. Anal. Chem. 62(19)2145-2148.

ASTM (American Society of Testing and Materials). 2006. Standard Test Method for Measuring the Toxicity of Sediment-Associated Contaminants with Estuarine and Marine Invertebrates. ASTM E1367-03. ASTM Annual Book of ASTM Standards, Vol. 11.06. American Society of Testing and Materials, West Conshohocken, PA.

Barbour, M.T., J. Gerritsen, B.D. Snyder, and J.B. Stribling. 1999. Rapid Bioassessment Protocols for Use in Streams and Wadeable Rivers: Periphyton, Benthic Macroinvertebrates and Fish, 2nd Ed. EPA 841-B-99-002. Office of Water, U.S. Environmental Protection Agency, Washington, DC [online]. Available: http://www.epa.gov/owow/monitoring/rbp/ [accessed Jan. 16, 2007].

Baumann, P.C. 2000. Health of Bullhead in an Urban Fishery after Remedial Dredging Final Report - January 31, 2000. Prepared for U.S. Environmental Protection Agency, Great Lakes National Program Office, Chicago IL, by U.S. Geological Survey Field Research Station, Ohio State University, Columbus, OH [online]. Available: http://www.epa.gov/glnpo/sediment/Bullhead/index.html [accessed Jan. 9, 2006].

BBL (Blasland, Bouck & Lee, Inc.). 1995. Non-Time-Critical Removal Action: Documentation Report, Vol. 1. Grasse River Study Area, Massena, New York, Aluminum Company of America, Massena, New York. Draft. Blasland, Bouck & Lee, Inc., Syracuse, New York. December 1995.

Bergen, B.J., W.G. Nelson, J. Mackay, D. Dickerson, and S. Jayaraman. 2005. Environmental monitoring of remedial dredging at the New Bedford Harbor, MA, Superfund site. J. Environ. Monit. Assess. 111(1-3):257-275.

Boese, B.L., H. Lee II, D.T. Specht, R.C. Randall, and M.H. Winsor. 1990. Comparison of aqueous and solid-phase uptake for hexachlorobenzene in the tellinid clam, *Macoma nasuta* (Conrad): A mass balance approach. Environ. Toxicol. Chem. 9(2):221-231.

Bohlen, W.F., and M.J. Erickson. 2006. Incorporating sediment stability within the management of contaminated sediment sites: A synthesis approach. Integr. Environ. Assess. Manage. 2(1):24–28.

Booij, K., H.M. Sleiderink, and F. Smedes. 1998. Calibrating the uptake kinetics of semipermeable membrane devices using exposure standards. Environ. Toxicol. Chem. 17(7):1236-1245.

Burton, G.A., Jr. 2002. Sediment quality criteria in use around the world. Limnology 3(2):65-76.

Burton, G.A., Jr., and J.F. Nordstrom. 2004. An *in situ* toxicity identification evaluation method. Part II. Field validation. Environ. Toxicol. Chem. 23(12):2851-2855.

Burton, G.A., Jr., and R. Pitt. 2002. Stormwater Effects Handbook: A Tool Box for Watershed Managers, Scientists and Engineers. Boca Raton, FL: Lewis [online]. Available: http://www.epa.gov/ednnrmrl/publications/books/handbook/index.htm [accessed Jan. 16, 2007].

Burton, G.A., Jr. and C. Rowland. 1998. Assessment of Sediment Toxicity in the Black River Watershed. Final Report. Great Lakes National Program Office, U.S. Environmental Protection Agency, Chicago, IL [online]. Available: http://epa.gov/greatlakes/sediment/BlackRiver/Index.htm [accessed Jan. 16, 2007].

Burton, G.A., Jr., R. Pitt, and S. Clark. 2000. The role of traditional and novel toxicity test methods in assessing stormwater and sediment contamination. CRC Crit. Rev. Environ. Sci. Technol. 30(4):413-447.

Burton, G.A., Jr., P.M. Chapman, and E.P. Smith. 2002. Weight-of-evidence approaches for assessing ecosystem impairment. Hum. Ecol. Risk Assess. 8(7):1675-1696.

Burton, G.A., Jr., M.S. Greenberg, C.D. Rowland, C.A. Irvine, D.R. Lavoie, J.A. Brooker, L. Moore, D.F.N. Raymer, and R.A. McWilliam. 2005. *In situ* exposures using caged organisms: A multi-compartment approach to detect aquatic toxicity and bioaccumulation. Environ. Pollut.134(1):133-144.

Chappie, D.J., and G.A. Burton, Jr. 2000. Applications of aquatic and sediment toxicity testing in situ. J. Soil Sediment. Contam. 9(3):219-246.

Choi, A.L., J.I. Levy, D.W. Dockery, L.M. Ryan, P.E. Tolbert, L.M. Altshul, and S.A. Korrick. 2006. Does living near a Superfund site contribute to higher polychlorinated biphenyl (PCB) exposure? Environ. Health Perspect. 114(7):1092-1098.

Clark, J.L., and W.H. Clements. 2006. The use of *in situ* and stream microcosm experiments to assess population and community-level responses to metals. Environ. Toxicol. Chem. 25(9):2306-2312.

Clements, W.H. 1999. Metal tolerance and predator-prey interactions in benthic macroinvertebrate stream communities. Ecol. Appl. 9(3):1073-1084.

Clements, W.H., and M.C. Newman. 2002. Community Ecotoxicology. Chichester, UK: Wiley.

Conder, J.M., and T.W. La Point. 2005. Solid phase microextraction for predicting the bioavailability of 2,4,6-trinitrotoluene and its primary transformation products in sediment and water. Environ. Toxicol. Chem. 24(5): 1059-1066.

Connolly, J.P., and R.V. Thomann. 1992. Modeling the accumulation of organic chemicals in aquatic food chains. Pp. 385-406 in Fate of Pesticides and Chemicals in the Environment, J.L. Schnoor, ed. New York: Wiley.

Crane, M., G.A. Burton, J.M. Culp, M.S. Greenberg, K.R. Munkittrick, R. Ribeiro, M.H. Salazar, and S.D. St-Jean 2007. Review of aquatic *in situ* approaches for stressor and effect diagnosis. Integr. Environ. Assess. Manage. 3(2):234-245.

Cressie, N. 1991. Statistics for Spatial Data. New York: Wiley.

Cullen, A.C., D.J. Vorhees, and L.M. Altshul. 1996. Influence of harbor contamination on the level and composition of polychlorinated biphenyls in produce in Greater New Bedford, Massachusetts. Environ. Sci. Technol. 30(5):1581-1588.

Dalton, Olmsted & Fuglevand, Inc. 2006. Remediation Action Construction Report, Part 1: Head of Hylebos Waterway Problem Area Commencement Bay Nearshore/Tideflats Superfund Site Tacoma, Washington, Review Draft, July 21, 2006. Prepared for Head of Hylebos Cleanup Group, Arkema, Inc, General Metals of Tacoma, Inc, by Dalton, Olmsted & Fuglevand, Inc., Kirkland, WA. July 21, 2006.

DeWitt, T.H., C.W. Hickey, D.J. Morrisey, M.G. Nipper, D.S. Roper, R.B. Williamson, L. Van Dam, and E.K. Williams. 1999. Do amphipods have the same concentration-response to contaminated sediment *in situ* as in vitro? Environ. Toxicol. Chem. 18(5):1026-1037.

Diggle, P.J., and P.J. Ribeiro, Jr. 2007. Model-based Geostatistics. New York: Springer.

Di Toro, D.M., J.D. Mahony, D.J. Hansen, K.J. Scott, M.B. Hicks, S.M. Mayr, and M.S. Redmond. 1990. Toxicity of cadmium in sediments: The role of acid volatile sulfide. Environ. Toxicol. Chem. 9(12):1487-1502.

Di Toro, D.M., C.S. Zarba, D.J. Hansen, W.J. Berry, R.C. Swartz, C.E. Cowan, S.P. Pavlou, H.E. Allen, N.A. Thomas, and P.R. Paquin. 1991. Technical basis for establishing sediment quality criteria for nonionic organic chemicals using equilibrium partitioning. Environ. Toxicol. Chem. 10(12):1541-1584.

Di Toro, D.M., J.D. Mahony, D.J. Hansen, K.J. Scott, A.R. Carlson, and G.T. Ankley. 1992. Acid volatile sulfide predicts the acute toxicity of cadmium and nickel in sediments. Environ. Sci. Technol. 26(1): 96-101.

Dow (Dow Chemical Company). 2006. GeoMorph Sampling and Analysis Plan: Upper Tittabawassee River, The Dow Chemical Company, Michigan Operations, Midland, MI [online]. Available: http://www.deq.state. mius/documents/deq-whm-hw-dow-Text-Final%20GeoMorphSAP-612006.pdf [accessed Jan. 17, 2007].

Elzinga, C.L., D.W. Salzer, and J.W. Willoughby. 1998. Measuring and Monitoring Plant Populations. BLM Technical Reference 1730-1. BLM/RS/ST-98/005+1730. Bureau of Land Management [online]. Available: http://www.blm.gov/nstc/library/pdf/MeasAndMon.pdf [accessed Jan. 17, 2007].

Environment Canada. 1992. Biological Test Method: Acute Test for Sediment Toxicity Using Marine or Estuarine Amphipods. Report EPS 1/RM/26. Ottawa (ON): Environment Canada. Conservation and Protection [on-line]. Available: http://www.etc-cte.ec.gc.ca/organization/bmd/pubs/pubs_en/1RM26Englishfinal.pdf [accessed May 11, 2007].

EPA (U.S. Environmental Protection Agency). 1994. Methods for Assessing the Toxicity of Sediment-Associated Contaminants with Estuarine and Marine Invertebrates. EPA 600/R-94/025. Office of Research and Development, U.S. Environmental Protection Agency, Washington, DC. June 1994.

EPA (U.S. Environmental Protection Agency). 1997. Report on the Effects of the Hot Spot Dredging Operations: New Bedford Harbor Superfund Site, New Bedford, MA. U.S. Environmental Protection Agency, Region 1. October 1997 [online]. Available: http://www.epa.gov/region01/superfund/sites/newbedford/47203.pdf [accessed Jan. 18, 2007].

EPA (U.S. Environmental Protection Agency). 1998. Guidelines for Ecological Risk Assessment. EPA/630/R095/002F. Risk Assessment Forum, U.S. Environmental Protection, Washington, DC. April 1998 [online]. Available: http://cfpub.epa.gov/ncea/cfm/recordisplay.cfm?deid=12460 [accessed Jan. 18, 2007].

EPA (U.S. Environmental Protection Agency). 1999. Evaluation of Toxicity and Bioaccumulation of Contaminants in Sediments Samples from Waukegan Harbor, Illinois. EPA-905-R-99-009. Great Lakes National Program Office, U.S. Environmental Protection Agency, Chicago, IL. October 1999 [online]. Available: http://www.epa.gov/glnpo/sediment/waukegan/whrpt.pdf [accessed Jan. 9, 2007].

EPA (U.S. Environmental Protection Agency). 2000a. Methods for Measuring the Toxicity and Bioaccumulation of Sediment-Associated Contaminants with Freshwater Invertebrates, 2nd Ed. EPA/600/R-99/064. Office of Research and Development, U.S. Environmental Protection Agency, Duluth, MI, and Office of Science and Technology, U.S. Environmental Protection Agency, Washington, DC. March 2000 [online]. Available: http://www.epa.gov/waterscience/cs/freshmanual.pdf [accessed May 11, 2007].

EPA (U.S. Environmental Protection Agency). 2000b. Equilibrium Partitioning Sediment Guidelines (ESGs) for the Protection of Benthic Organisms: Metal Mixtures (Cadmium, Copper, Lead, Nickel, Silver and Zinc). EPA 822R02045. Office of Science and Technology, U.S Environmental Protection Agency, Washington, DC.

EPA (U.S. Environmental Protection Agency). 2000c. Guidance for Assessing Chemical Contaminant Data for Use in Fish Advisories, Vols. 1 and 2. EPA 823-B-00-007. EPA 823-B-00-008. Office of Water, U.S. Environmental Protection Agency, Washington, DC [online]. Available: http://www.epa. gov/waterscience/fish/guidance.html [accessed Jan. 18, 2007].

EPA (U.S. Environmental Protection Agency). 2000d. Guidance for Data Quality Assessment: Practical Methods for Data Analysis, EPA QA/G9, QA 00/Update. EPA/600/R-96/084. Office of Environmental Information, U.S. Environmental Protection Agency, Washington, DC [online]. Available: http://www.epa.gov/quality/qs-docs/g9-final.pdf [accessed Jan. 18, 2007].

EPA (U.S. Environmental Protection Agency). 2001a. Methods for Collection, Storage and Manipulation of Sediments for Chemical and Toxicological Analyses: Technical Manual. EPA-823-B-01-002. Office of Water, U.S. Environmental Protection Agency, Washington DC. October 2001 [online]. Available: http://www.epa.gov/waterscience/cs/toc.pdf [accessed Jan. 18, 2007].

EPA (U.S. Environmental Protection Agency). 2001b. Comprehensive Five-Year Review Guidance. EPA 540-R-01-007. Office of Emergency and Remedial Response, U.S. Environmental Protection Agency, Washington, DC. June 2001 [online]. Available: http://www.epa.gov/superfund/resources/5year/ guidance.pdf [accessed Aug. 3, 2006].

EPA (U.S. Environmental Protection Agency). 2003a. Procedures for the Derivation of Equilibrium Partitioning Sediment Benchmarks (ESBs) for the Protection of Benthic Organisms: Dieldrin. EPA/600/R-02/010. Office of Research and Development, U.S Environmental Protection Agency, Washington, DC [online]. Available: http://www.epa.gov/nheerl/public ations/files/dieldrin.pdf [accessed May 11, 2007].

EPA (U.S. Environmental Protection Agency). 2003b. Procedures for the Derivation of Equilibrium Partitioning Sediment Benchmarks (ESBs) for the Protection of Benthic Organisms: Endrin. EPA/600/R-02/009. Office of Research and Development,U.S Environmental Protection Agency, Washington, DC [online]. Available: http://www.epa.gov/nheerl/publications/files/endrin. pdf [accessed May 11, 2007]. August 2003.

EPA (U.S. Environmental Protection Agency). 2003c. Procedures for the Derivation of Equilibrium Partitioning Sediment Benchmarks (ESBs) for the Protection of Benthic Organisms: PAH Mixtures EPA/600/R-02/013. Office of Research and Development, U.S Environmental Protection Agency, Washington, DC [online]. Available: http://www.epa.gov/nheerl/publications/ files/PAHESB.pdf [accessed May 11, 2007].

EPA (U.S. Environmental protection Agency). 2004. Guidance for Monitoring at Hazardous Waste Sites: Framework for Monitoring Plan Development and Implementation. OSWER Directive 9355.4-28. Office of Solid Waste and

Emergency Response, U.S. Environmental Protection Agency. January 2004 [online]. Available: http://www.epa.gov/superfund/action/guidance/dir 9355.pdf [accessed Jan. 18, 2007].

EPA (U.S. Environmental Protection Agency). 2005a. Contaminated Sediment Remediation Guidance for Hazardous Waste Sites. EPA-540-R-05-012. OSWER 9355.0-85. Office of Solid Waste and Emergency Response, U.S. Environmental Protection Agency. December 2005 [online]. Available: http://www.epa.gov/superfund/resources/sediment/pdfs/guidance.pdf [accessed Dec. 26, 2006].

EPA (U.S. Environmental Protection Agency). 2005b. Procedures for the Derivation of Equilibrium Partitioning Sediment Benchmarks (ESBs) for the Protection of Benthic Organisms: Metal Mixtures (Cadmium, Copper, Lead, Nickel, Silver and Zinc). EPA 600/R-02/011. Office of Research and Development, U.S Environmental Protection Agency, Washington, DC. January 2005 [online]. Available: http://www.epa.gov/nheerl/publications/files/metalsESB_022405.pdf [accessed May 11, 2007].

EPA/USACE (U.S. Environmental Protection Agency and U.S. Army Corps of Engineers). 1998. Evaluation of Dredged Material Proposed for Discharge in Waters of the U.S.—Testing Manual. EPA-823 -B-98 -004. U.S. Environmental Protection Agency, and Department of Army, U.S. Army Corps of Engineers. February 1998 [online]. Available: http://www.epa.gov/water science/itm/ITM/ [accessed Jan. 18, 2007].

Fitzgerald, E.F., K.A. Brix, D.A. Deres, S.A. Hwang, B. Bush, G. Lambert, and A. Tarbell. 1996. Polychlorinated biphenyl (PCB) and dichlorodiphenyl dichloroethylene (DDE) exposure among Native American men from contaminated Great Lakes fish and wildlife. Toxicol. Ind. Health 12(3-4):361–368.

Fitzgerald, E.F., D.A. Deres, S.A. Hwang, B. Bush, B.Z. Yang, A. Tarbell, and A Jacobs. 1999. Local fish consumption and serum PCB concentrations among Mohawk men at Akwesasne. Environ. Res. 80(2Pt.2):97–103.

Fitzgerald, E.F., S.A. Hwang, K. Langguth, M. Cayo, B.Z. Yang, B. Bush, P. Worswick, and T. Lauzon. 2004. Fish consumption and other environmental exposures and their associations with serum PCB concentrations among Mohawk women at Akwesasne. Environ. Res. 94(2):160–170.

Fitzgerald, E.F., S.A. Hwang, G. Lambert, M. Gomez, and A. Tarbell. 2005. PCB Exposure and in vivo CYP1A2 activity among Native Americans. Environ. Health Perspect. 113(3):272-277.

Fredette, T.J., D.A. Nelson, T. Miller-Way, J.A. Adair, V.A. Sotler, J.E. Clausner, E.B. Hands, and F.J. Anders. 1990. Selected Tools and Techniques for Physical and Biological Monitoring of Aquatic Dredged Material Disposal Sites. Technical Report D-90-11. U.S. Army Corps of Engineers, Waterways Experiment Station, Vicksburg, MS [online]. Available: http://stinet.dtic.

mil/oai/oai?&verb=getRecord&metadataPrefix=html&identifier=ADA22944 2 [accessed Jan. 17, 2007].

Goovaerts, P. 1997. Geostatistics for Natural Resources Evaluation. New York: Oxford University Press.

Greenberg, M.S., G.A. Burton, Jr., and C.D. Rowland. 2002. Optimizing interpretation of *in situ* effects of riverine pollutants: Impact of upwelling and downwelling. Environ. Toxicol. Chem. 21(2):289–297.

Hart, D.D. 1994. Building a stronger partnership between ecological research and biological monitoring. J. N. Am. Benthol. Soc. 13(1):110-116.

Hofelt, C.S., and D. Shea. 1997. Accumulation of organochlorine pesticides and PCBs by semipermeable membrane devices and *Mytilus edulis* in New Bedford Harbor. Environ. Sci. Technol. 31(1):154 – 159.

Huckins, J.N., M.W. Tubergen, and G.K. Manuweera. 1990. Semipermeable membrane devices containing model lipid: A new approach to monitoring the bioavaiiability of lipophilic contaminants and estimating their bioconcentration potential. Chemosphere 20(5):533-552.

Huckins, J.N., G.K. Manuweera, J.D. Petty, D. Mackay, and J.A. Lebo. 1993. Lipid-containing semipermeable membrane devices for monitoring organic contaminants in water. Environ. Sci. Technol. 27(12):2489-2496.

Ingersoll, C.G., P.S. Haverland, E.L. Brunson, T.J. Canfield, F.J. Dwyer, C.E. Henke, N.E. Kemble, D.R. Mount and R.G. Fox. 1996. Calculations and evaluation of sediment effect concentrations for the amphipod *Hyalella azteca* and the midge *Chironomus riparius*. J. Great Lakes Res. 22(3):602-623.

Ingersoll, C.G., S.M. Bay, J.L. Crane, L.J. Field, T.H. Gries, J.L. Hyland, E.R. Long, D.D. MacDonald, and T.P. O'Connor. 2005. Ability of SQGs to estimate effects of sediment-associated contaminants in laboratory toxicity tests or in benthic community assessments. Pp. 497-556 in Use of Sediment Quality Guidelines and Related Tools for the Assessment of Contaminated Sediments, R.J. Wenning, G.E. Batley, C.G. Ingersoll, and D.W. Moore, eds. Pensacola, FL: SETAC Press.

Ireland, D.S., G.A. Burton, Jr., and G.G. Hess. 1996. *In Situ* toxicity evaluations of turbidity and photoinduction of polycyclic aromatic hydrocarbons. Environ. Toxicol. Chem. 15(4):574-581.

Jager, T., F.A. Antón Sánchez, B. Muijs, E.G. van der Velde, and L. Posthuma. 2000. Toxicokinetics of polycyclic aromatic hydrocarbons in *Eisenia andrei* (Oligochaeta) using spiked soil. Environ. Toxicol. Chem. 19(4):953-961.

Johnson, J.C., and J.M. van Emon. 1996. Quantitative enzyme-linked immunosorbent assay for determination of polychlorinated biphenyls in environmental soil and sediment samples. Anal. Chem. 68 (1):162-169.

Karr, J.R. 1993. Defining and assessing ecological integrity: Beyond water quality. Environ. Toxicol. Chem. 12(9):1521-1531.

Kelaher, B.P., J.S. Levinton, J. Oomen, B.J. Allen, and W.H. Wong. 2003. Changes in benthos following the clean-up of a severely metal-polluted cove in the Hudson River Estuary: Environmental restoration or ecological disturbance? Estuaries 26(6):1505-1516.

Kemble, N.E., D.G. Hardesty, C.G. Ingersoll, B.T. Johnson, F.J. Dwyer, and D.D. MacDonald. 2000. An evaluation of the toxicity of contaminated sediments from Waukegan Harbor, Illinois, following remediation. Arch. Environ. Contam. Toxicol. 39(4):452-461.

Korrick, S.A., and L. Altshul. 1998. High breast milk levels of polychlorinated biphenyls (PCBs) among four women living adjacent to a PCB-contaminated waste site. Environ. Health Perspect. 106(8):513-518.

Kraaij, R., P. Mayer, F.J. Busser, M. van het Bolscher, W. Seinen, J. Tolls, and A.C. Belfroid. 2003. Measured pore-water concentrations make equilibrium partitioning work: A data analysis. Environ. Sci. Technol. 37(2):268-274.

Lanno, R.P., J. Wells, J. Conder, K. Bradham, and N. Basta. 2004. The bioavailability of chemicals in soil for earthworms. Ecotox. Environ. Safe. 57(1):39-47.

Lanno, R.P., T.W. La Point, J.M. Conder, and J.B. Wells. 2005. Applications of solid-phase microextraction fibers as biomimetic sampling devices in ecotoxicology. Pp. 511-523 in Techniques in Aquatic Toxicology, Vol. 2, G. Ostrander, ed. Boca Raton: CRC Press.

Leslie, H.A., T.L. ter Laak, F.J.M. Busser, M.H.S. Kraak, and J.L.M. Hermens. 2002. Bioconcentration of organic chemicals: Is a solid-phase microextraction fiber a good surrogate for biota? Environ. Sci. Technol. 36(24):5399-5404.

Levinton, J.S., E. Suatoni, W. Wallace, R. Junkins, B. Kelaher, and B.J. Allen. 2003. Rapid loss of genetically based resistance to metals after the cleanup of a Superfund site. Proc. Natl. Acad. Sci. U.S.A 100(17):9889-9891.

Linkov, I., D. Burmistrov, J. Cura, and T.S. Bridges. 2002. Risk-based management of contaminated sediments: Consideration of spatial and temporal patterns in exposure modeling. Environ. Sci. Technol. 36(2):238-246.

Lohmann, R., R.M. Burgess, M.G. Cantwell, S.A. Ryba, J.K. MacFarlane, P.M. Gschwend. 2004. Dependency of polychlorinated biphenyl and polycyclic aromatic hydrocarbon bioaccumulation in Mya Arenaria on both water column and sediment bed chemical activities. Environ. Toxicol. Chem. 23(11):2551-2562.

Long, E.R., and L.G. Morgan. 1990. The Potential for Biological Effects of Sediment-Sorbed Contaminants Tested in the National Status and Trends Program. NOAA Technical Memorandum NOS OMA 52. National Oceanographic and Atmospheric Administration, Seattle, WA. March 1990 [online]. Available: http://www.ccma.nos.noaa.gov/publications/tm52.pdf [accessed May 11, 2007].

Long, E.R., and D.D. MacDonald. 1998. Recommended uses of empirically derived sediment quality guidelines for marine and estuarine ecosystems. Hum Ecol Risk Assess. 4:1019-1039.

Lu, X., D.D. Reible, and J.W. Fleeger. 2004. Relative importance of ingested sediment versus pore water as uptake routes for PAHs to the deposit-feeding oligochaete *Ilyodrilus templetoni*. Arch. Environ. Contam. Toxicol. 47(2):207-214.

Lu, X., D.D. Reible, and J.W. Fleeger. 2006. Bioavailability of polycyclic aromatic hydrocarbons in field-contaminated Anacostia River (Washington, DC) sediment. Environ. Toxicol. Chem. 25(11):2869-2874.

Luoma, S.N., and P.S. Rainbow. 2005. Why is metal bioaccumulation so variable? Biodynamics as a unifying concept. Environ. Sci. Technol. 39(7):1921-1931.

MacDonald, D.D., C.G. Ingersoll, and T.A. Berger. 2000. Development and evaluation of consensus-based sediment quality guidelines for freshwater ecosystems. Arch. Environ. Contam. Toxicol. 39(1):20–31.

Mason, R.L., R.F. Gunst, and J.L. Hess. 1989. Statistical Design and Analysis of Experiments with Applications to Engineering and Science. New York: Wiley.

McDonald, L.L., S. Howlin, J. Polyakova, and C.J. Bilbrough. 2003. Evaluation and Comparison of Hypothesis Testing Techniques for Bond Release Applications. Prepared for Abandoned Coal Mine Lands Research Program, University of Wyoming, Laramie, WY, by Western EcoSystems Technology, Inc., Cheyenne, WY, and Land Quality Division, Department of Environmental Quality, Cheyenne, WY. May 28, 2003 [online]. Available: http://www.west-inc.com/reports/aml.pdf [accessed Jan.18, 2007].

MA DPH (Massachusetts Department of Public Health). 1997. Housatonic River Area PCB Exposure Assessment Study, Final Report. September 1997. Bureau of Environmental Health Assessment, Environmental Toxicology Unit, Massachusetts Department of Public Health.

McLeod, P.B., M.J. van den Heuvel-Grave, S.N. Luoma, and R.G. Luthy. 2007. Biological uptake of polychlorinated biphenyls by *Macoma balthica* from sediment amended with activated carbon. Environ. Toxicol. Chem. 26(5): 980-987.

Miller, D.T., S.K. Condon, S. Kutzner, D.L. Phillips, E. Krueger, R. Timperi, V.W. Burse, J. Cutler, and D.M. Gute. 1991. Human exposure to polychlorinated biphenyls in greater New Bedford, Massachusetts: A prevalence study. Arch. Environ. Contamin. Toxicol. 20(3):410-416.

Nelson, W. 2006. Long-term chemical and biological monitoring efforts at the New Bedford Harbor (NBH) Superfund site. Presentation at the Second Meeting on Sediment Dredging at Superfund Megasites, July 12, 2006, New Bedford, MA.

Nichols, F.H., and J.K. Thompson. 1985. Time scale of change in the San Francisco Bay benthos. Hydrobiologia 129(1):121-138.

Norstrom, R.J., A.E. McKinnon, and A.S.W. DeFreitas. 1976. A bioenergetics-based model for pollutant accumulation by fish. Simulation of PCB and methylmercury residue levels in Ottawa River yellow perch (*Perca flavescens*). J. Fish. Res. Board Can. 33(2):248-267.

NRC (National Research Council). 2001. A Risk-Management Strategy for PCB-Contaminated Sediments. Washington, DC: National Academy Press.

NRC (National Research Council). 2003. Bioavailability of Contaminants in Soils and Sediments. Washington, DC: National Academies Press.

Palermo, M. 2006. Overview of Dredging Equipment, Processes: Operating Methods and Strategies. Presentation at the First Meeting on Sediment Dredging at Superfund Megasites, March 22, 2006, Washington DC.

Salazar, M.H., and S.M. Salazar. 1997. Using caged bivalves to characterize exposure and effects associated with pulp and paper mill effluents. Water Sci. Technol. 35(2-3):213-220.

Shaw Environmental, Inc. 2006. Final Basis of Design Report: Lower Fox River and Green Bay Site, Brown, Outagamie, and Winnebago Counties, Wisconsin. Prepared for Fort James Operating Company, Inc., NCR Corporation, by Shaw Environmental, Inc. June 16, 2006 [online]. Available: http://www.dnr.state.wi.us/org/water/wm/foxriver/documents/BODR/Final_BODR_VolumeI.pdf [accessed Jan 18, 2007].

Solomon, K.R., G.T. Ankley, R. Baudo, G.A. Burton, Jr., C.G. Ingersoll, W. Lick, S.N Luoma, D.D. MacDonald, T.B. Reynoldson, R.C. Swartz, and W. Warren-Hicks. 1997. Workgroup summary report on methodological uncertainty in conducting sediment ecological risk assessments with contaminated sediments. Pp. 271-296 in Ecological Risk Assessments of Contaminated Sediments, C.G. Ingersoll, T. Dillon, and G. R. Biddinger, eds. Pensacola, FL: SETAC Press.

Ta, N. 2001. Rapid Sediment Characterization of PCBs with ELISA: An Immunoassay Technique a Rapid Sediment Characterization (RSC) Tool. TechData Sheet. TDS-2086-ENV. Naval Facilities Engineering Service Center, Port Hueneme, CA. June 2001 [online]. Available: http://costperformance.org/monitoring/pdf/elisa_2.pdf [accessed Jan. 18, 2007].

Tucker, K.A., and G.A. Burton, Jr. 1999. Assessment of nonpoint source runoff in a stream using *in situ* and laboratory approaches. Environ. Toxicol. Chem. 18(12):2797-2803.

Vinturella, A.E., R.M. Burgess, B.A. Coull, K.M. Thompson, and J.P. Shine. 2004. Use of passive samplers to mimic uptake of polycyclic aromatic hydrocarbons by benthic polychaetes. Environ. Sci. Technol. 38(4):1154-1160.

Viollier, E., C Rabouille, S.E. Apitz, E. Breuer, G. Chaillou, K. Dedieu, Y. Furakawa, C. Grentz, P. Hall, F. Janssen, J.L. Morford, J-C. Poggiale, S. Roberts,

G.T. Shimmield, M. Taillefert, A. Tengberg, F. Wenzhoefer, and U. Witte. 2003. Benthic biogeochemistry: State of the art technologies and guideline for the future of *in situ* survey. J. Exp. Mar. Biol. Ecol. 285:5-31.

Weis, J.S., and P. Weis. 1989. Tolerance and stress in a polluted environment: The case of mummichog. BioScience 39(2):89-95.

Wells, J.B., and R.P. Lanno. 2001. Passive Sampling Devices (PSDs) as biological surrogates for estimating the bioavailability of organic chemicals in soil. Pp. 253-270 in Environmental Toxicology and Risk Assessment: Science, Policy, and Standardization – Implications for Environmental Decisions: Tenth Volume, B.M. Greenberg, R.N. Hull, M.H. Roberts, Jr., and R.W. Gensemer, eds. ASTM STP 1403. West Conshohocken, PA: American Society for Testing and Materials.

Wenning, R.J., and C.G. Ingersoll. 2002. Summary of the SETAC Pellston Workshop on Use of Sediment Quality Guidelines and Related Tools for the Assessment of Contaminated Sediments, 17-22 August 2002; Fairmont, MT. Society of Environmental Toxicology and Chemistry (SETAC), Pensacola FL [online]. Available: http://www.cerc.usgs.gov/pubs/sedtox/WG0_SETAC_SQG_Summary.pdf. [accessed May 11, 2007].

Wenning, R.J., G.E., Batley, C.G. Ingersoll, and D.W. Moore, eds. 2005. Use of Sediment Quality Guidelines and Related Tool for the Assessment of Contaminated Sediments. Pensacola. FL: Society of Environmental Toxicology & Chemistry Press. 783 pp.

Weston, D.P., D.L. Penry, and L.K. Gulmann. 2000. The role of ingestion as a route of contaminant bioaccumulation in a deposit-feeding polychaete. Arch. Environ. Contam. Toxicol. 38(4):446-454

Weston (Weston Solutions Inc.). 2005. Final Manistique Harbor and River Site, Manistique, Michigan, Data Evaluation Report: Revision 1-19 May 2005. Document No. RWW236-2A-ATGN. Prepared for U.S. Environmental Protection Agency, Chicago, IL.

Wharfe, J., W. Adams, S.E. Apitz, R. Barra, T.S. Bridges, C. Hickey, and S. Ireland. 2007. *In situ* methods of measurement-an important line of evidence in the environmental risk framework. Integr Environ. Assess. Manage. 3(2):268-274.

Word, J.Q., B.B. Albrecht, M.L.Anghera, R. Baudo, S.M. Bay, D.M. Di Toro, J.L. Hyland, C.G. Ingersoll, P.F. Landrum, E.R. Long, J.P. Meador, D.W. Moore, T.P. O'Connor, and J.P. Shine. 2005. Predictive ability of sediment quality guidelines. Pp. 121-161 in Use of Sediment Quality Guidelines and Related Tools for the Assessment of Contaminated Sediments, R.J. Wenning, G.E. Batley, C.G. Ingersoll, and D.W. Moore, eds. Pensacola, FL: SETAC Press).

You, J., P.F. Landrum, and M.J. Lydy. 2006. Comparison of chemical approaches for assessing bioavailability of sediment-associated contaminants. Environ. Sci. Technol. 40(20):6348-6353.

Yount, J.D., and G.J. Niemi. 1990. Recovery of lotic communities and ecosystems from disturbance: A narrative review of case studies. Environ. Manage. 14(5):547-569.

Zhang, H., W. Davison, B. Knight, and S. McGrath. 1998. *In situ* measurements of solution concentrations and fluxes of trace metals in soils using DGT. Environ. Sci. Technol. 32(5):704-710.

6

Dredging at Superfund Megasites: Improving Future Decision-Making

INTRODUCTION

The preceding chapters discussed sediment management and dredging at Superfund megasites and included sections on assessing the effectiveness of dredging for removing contaminated sediment to attain remedial-action objectives and achieve specified cleanup levels. The assessment included the review of 26 projects from which general conclusions were developed with respect to the appropriate use and limitations of dredging in meeting risk-based goals. From those conclusions, the committee developed guidelines with respect to favorable site conditions under which dredging should be more likely to achieve long-term remedial-action objectives. The committee also offered recommendations for monitoring to facilitate scientifically based and timely decision- making to improve dredging effectiveness.

In this final chapter, the committee addresses the charge in the statement of task to consider "how conclusions about completed and current operations can inform future remedial decision-making" and to "develop recommendations that will facilitate scientifically based and timely decision making for megasites in the future." Specifically, we seek to identify how lessons learned from experience may inform future practices and management of contaminated sediment at megasites. This includes the expected role of dredging in the future and the issues and factors that need to be addressed to ensure the effective use of dredging as a component of contaminated sediment remediation. Most of the committee's earlier recommendations focus on these issues at the site-specific level, but this chapter focuses on the national level.

MANAGING SEDIMENT MEGASITES IN THE FUTURE

With the establishment of Superfund in 1980, we now have the opportunity for retrospective analysis at dozens of sediment sites to evaluate decision-making, field experience, and remedial effectiveness where dredging has been selected for sediment cleanup. In the past, a rigorous evaluation of whether site remediation achieved risk reduction goals and what factors contributed to or limited the achievement of those goals was often just not done. Although information is available from various sites with respect to volume of bulk material removed or sediment concentration achieved, that information does not permit determination of the degree to which remedial objectives for risk-reduction were achieved. Thus, it is not easy to determine which approaches resulted in risk reduction under various site conditions. The difficulty stems partly from the lack of comprehensive post-dredging monitoring data and from the fact that followup assessments typically do not quantify uncertainty in both risk measurements and predictions.

In hindsight, it is clear that there are limitations to dredging effectiveness. With this historical perspective comes the opportunity to learn and improve how we think about and implement environmental dredging. Perhaps nothing is more important than to step back and derive common lessons from experience, as was done in Chapter 4. This type of review needs to be continuous and needs to part of a shared experience among regulators, practitioners, and the public.

As described in Chapter 2, sediment megasites are among the most challenging and costly sites on the National Priorities List (NPL). Megasites are conventionally defined as sites with remedial activities costing at least $50 million; the U.S. Environmental Protection Agency (EPA) has defined contaminated sediment megasites as sites for which the *sediment component* of remedial activities will cost at least $50 million.

The charge to this committee focused on megasites, but the dearth of such sites with completed dredging remedies and with good pre-dredging and post-dredging data has meant that the committee reviewed smaller sites, or individual projects at megasites. The projects evaluated did not include any of the magnitude (that is, tens of miles of river stretches and thousands of hectares) and time frames (for example, decades in the Hudson River [TAMS Consultants, 2000]) that can be anticipated at the largest of the current and future megasites. Megasites with a broad spatial area and large volume of contaminated sediments will likely require multiple seasons of dredging.[1] The larger scale and time frames will increase the chemical exposures and residual production related to operations, and make it more difficult to fully characterize contaminant distribution, sources of contamination, and conditions unfavorable for remedial operations. At these sites, risk based goals may not be achievable in the foreseeable future due to the long time frame, complexity of the sites, and the limitations of available technologies. The committee recognizes that experience with remediation at larger sites might reveal challenges not faced at the smaller sites and has attempted to anticipate such challenges to the extent possible in making its recommendations.

Cleanup of contaminated sediment megasites incorporates large temporal and spatial scales that create two distinct issues: the human health and ecosystem risk-reduction benefits achieved by isolated remediation in a large-scale watershed are difficult to predict and quantify; and the large spatial scales and long time lines, coupled with the complexity and heterogeneity of large-scale megasites, suggest that varied and combined remedial approaches will be appropriate. We can do a better job of addressing those issues by taking a broader, basin-wide view in contaminated sediment management and by embracing more flexible approaches. Those issues are discussed further below.

The Need for Regional-Scale Perspectives

Because contaminated sediment megasites are influenced by regional-scale phenomena, watershed and airshed contributions to sedi-

---

[1]For example, at current levels of operation, it will take more than 25 years to complete dredging at New Bedford (Dickerson and Brown 2006).

ment contamination at any site must be viewed in a larger framework to permit valid predictions about cleanup and risk reduction. Sediment megasites can span an entire waterbody (such as the Lower Fox River in Wisconsin) or be located in a watershed amongst other contaminated sediment sites. Bridges et al (2006) comment that several watersheds in the United States contain multiple contaminated sediment sites in close proximity to one another and that effective sediment management will require a more holistic approach to understanding multiple sources of contaminants (sediments, outfalls, and non-point sources) and their cumulative impacts in a waterway.

The public has a right to know what benefits will be achieved for particular investments. For example, if the risk being addressed is associated with the consumption of fish, a valid question is, How much will contaminants in fish decrease as a result of this action? Some organisms travel great distances and there is need to understand their movement and variable exposure to the Superfund sites. At the same time, the long-distance movement of toxics from the site throughout the larger water body needs to be understood. Finally, there needs to be an understanding of secular changes in basin-wide conditions and how they might relate to cleanup at a specific site. (For example, whether basin-wide concentrations of contaminants are declining because of point source reduction or whether contaminants are migrating in from other contaminated areas in the wider basin.) These factors contribute heavily in evaluating site specific data on concentrations in fish species and the broader water body before and after a cleanup.

As such, a regional approach to modeling and analysis of contaminants at megasites is needed to better understand their effect on resident and migratory fish and on the flux of contaminants within the wider basin (e.g. Linkov et al. 2002; von Stackelberg et al. 2002). Because of the difficulty in accurately estimating several of the necessary parameters and inputs for these types of models (particularly fish exposure to contaminated sediments, differences in movement of various fish populations or life stages, contaminant concentrations in prey, and uptake and loss kinetics of mobile species occupying areas of high and low exposure), their uncertainty will remain a concern. However, because these issues are of particular importance at megasites and where multiple Superfund sites exist in close proximity, the development of these models is essential.

A related issue is the lack of essential tools for understanding how reductions in sediment toxicity or biologic exposure will enhance ecosystem response and benefit ecosystem recovery. Much of our understanding of these topics is wholly observational or is derived from ancillary measures (such as sediment chemistry). In itself, such information provides little capability for predicting community or organism response after remediation. Understanding the ecosystem dynamics that affect recovery entails larger regional-scale phenomena, such as larval recruitment, food-web interactions, and fate and transport processes.[2]

The Use of Adaptive Management at Megasites

Given the difficulty in predicting dredging effectiveness, and the limited number of available alternative technologies, what changes can be made to improve the remedy selection and implementation process to ensure more effective and cost-effective remedies?

A major challenge to decision-makers is the uncertainty about whether—and how well—a remedy will work at a site. Experience has shown the wisdom of well-designed pilot field tests and experimentation prior to committing to a specific final cleanup remedy. Pilot testing, including monitoring of appropriate environmental variables, for example as part of the feasibility study, is used to test the performance of a technology or approach, understand the factors affecting its performance, and to provide information on how, if necessary, the remedy should be adapted to achieve desired goals. In this way, the information generated in the pilot tests and monitoring becomes a key component of the remedy selection, design, and implementation process.

The use of a structured process of selecting a management action, monitoring the effects of the action, and applying those lessons to optimize a management action is generally referred to as adaptive management (e.g., NRC 2003, 2004, 2005; Bridges et al. 2006; Linkov et al 2006a, b). As described in NRC 2004, "There is no prototype for its implementation, and no 'cookbook'-type set of steps or building blocks that will immediately constitute an adaptive-management program. It is context-specific, it involves feedback and learning between scientists, managers,

---

[2]As discussed in Chapter 5, biodynamic approaches that are linked with principles of functional ecology can help to bridge this knowledge gap.

and stakeholders." NRC (2005) recommends an adaptive-management approach at Superfund megasites where it is unlikely that final remedies can be identified and implemented. That report describes adaptive management as a six-step interactive process for defining and implementing management policies under conditions of high uncertainty regarding results of remedial actions (see Box 6-1).

Bridges et al. (2006) described the need for greater flexibility in sediment management processes because "the more strictly linear decision making process characterized as the 'decide and defend' approach to remedial decision making does not contain sufficient flexibility" and is unable to accommodate or benefit from other approaches such as adaptive management. In the current Superfund process, the ROD is the end result of a long and often difficult and contentious process of conducting studies and receiving and responding to input from stakeholders often with divergent and impassioned views of the type and extent of remediation that is required (see Chapter 2 for greater detail on the remedy selection process). A ROD often selects a specific remedy and predicts its ability to achieve cleanup levels and remedial action objectives at the site. Because the scale of megasites is so large, a variety of unanticipated conditions can greatly influence the results of a remediation. When remedies are selected without the benefit of actual, on-the-ground feedback on the effect of the remediation, there is a greater chance that unforeseen conditions and events will hinder progress or limit the effectiveness of the remediation. The ROD process can be reopened to amend or modify a ROD on the basis of information gathered after implementation begins, however, instituting an adaptive-management process from the outset recognizes the uncertainty inherent in predicting remedial results and allows adaptation of the remedy based on site experience to optimize progress toward attaining remedial goals.

In this process, the primary goals of Superfund, the protection of human health and the environment, remain paramount. As such, adaptive management is not a means to permit or sanction less rigorous cleanups, or to avoid public input or scrutiny of the decision making process. The principles of transparency and public notification remain essential and the adaptive-management process at a site needs to be developed in concert with stakeholders and insights from monitoring and testing need to be shared with them so that they can contribute to

---

BOX 6-1 Six-Step Adaptive-management Process

1.   Assessing the problem, including establishing measurable management objectives, key indicators of those objectives, quantitative or conceptual models to predict effects of remedial alternatives on the indicators, and forecasts of responses of indicators to remedial actions.

2.   Designing a management plan, including comparing and selecting remedial actions and, importantly, selecting indicator values that will trigger a change in management actions.

3.   Implementing the plan, including documenting and agreeing with stakeholders on those circumstances that might require deviations from the plan.

4.   Monitoring for effectiveness and for verifying and updating the conceptual model.

5.   Evaluating results obtained from monitoring, including comparing results with forecasts from earlier modeling, seeking to explain why results occurred, and provide recommendations for future action.

6.   Adjusting the management plan in response to the monitoring results, including implementing recommendations, reviewing and updating models, and developing new forecasts, management objectives, and management actions as necessary.

Source: Adapted from NRC 2005.

---

adapting the remedy, if necessary. It is expected that adaptive management could be implemented in the current legislative framework because CERCLA and the NCP have great flexibility and do not preclude adaptive management (NRC 2005). That implementation would need to be reviewed by EPA to best fit CERCLA requirements.

There is progress toward implementing adaptive management at contaminated sediment sites. Recent EPA guidance (EPA 2005) endorses the general concept, stating

> Project managers are encouraged to use an adaptive management approach, especially at complex sediment sites to provide additional certainty of information to support decisions. In general, this means testing of hypotheses and conclusions and reevaluating site assumptions as new information is gathered.

There are also examples of sites where adaptive-management principles, if not an explicit, rigorous adaptive-management process, have
been applied in remediating contaminated sediment sites (see Box 6-2).

In sum, the desired outcome of this more flexible approach is to allow, indeed to encourage, adaptation to realities on the ground in an effort to achieve remedial goals in as efficient and cost effective manner as
possible. These suggested changes reflect the need to make decisions in
the face of uncertainty while allowing managers and stakeholders to respond to, and take advantage of, unanticipated events and a variety of
possible future outcomes through the design of a flexible, iterative learning process.

BOX 6-2 Examples of the Application of Adaptive-management
Principles in Sediment Remediation

In the Fox River, WI, two demonstration projects were conducted during
the Remedial Investigation/Feasibility Study at Sediment Management Units
56/57 and Deposit N (Foth and Van Dyke, 2000; Montgomery Watson, 2001). The
projects provided useful information on implementability, effectiveness, and
expense of large-scale dredging at the site and were used to inform future decision-making.

In Operable Unit (OU) 1 of the Fox River, the ROD permits flexibility in
achieving cleanup levels and stipulates additional actions (further dredging or
capping) if dredging doesn't achieve desired results. Following dredging, if sampling shows that the 1 ppm action level has not been achieved, a surface-
weighted average concentration (SWAC) of 0.25 ppm may be used to assess the
effectiveness of PCB removal. If that SWAC of 0.25 ppm has not been achieved
for OU 1, the first option is that additional dredging may be undertaken to ensure that all sediments with PCB concentrations greater than the 1 ppm action
level are removed throughout the particular deposit. A second option is placing
a sand cover on dredged areas to reduce surficial concentrations to achieve a
SWAC of 0.25 ppm for OU 1 (WI DNR/EPA 2002).

Finally, in the case of the Grasse River, several large-scale dredging and
capping projects have revealed site-specific conditions that limited the effectiveness of the remediation (including dredging and capping). This site-specific information can then be used in development of a revised Analysis of Alternatives
Report (Alcoa Inc. 2005).

The Future of Dredging

While some improvements have been made to dredge design and operation (for example, precision positioning systems or dredge head or bucket modifications to reduce resuspension), in many respects dredging as a technology has not changed dramatically in the last few decades. What has changed is how dredging is applied. Devices designed and proven for navigational or maintenance dredging are now pressed into service for specific and precise contaminant-mass removal or to attain specific sediment contaminant concentrations in what are often complex settings and difficult conditions. In addition, it is often difficult to accurately characterize the sites, and define the degree of uncertainty about the effectiveness of different remedial approaches.

The committee found that most of the sites that it examined exhibited one or more conditions unfavorable for dredging and concluded that dredging alone is unlikely to be effective in achieving both short-term and long-term cleanup levels at many sites. However, its effectiveness as a contaminant-mass removal technology will ensure its use at most sites where mass removal is necessary (such as where navigational, source reduction, or sediment stability concerns are present). Where unfavorable conditions exist, it is likely to be implemented—in conjunction with capping, in situ treatment, or monitored natural recovery—as part of a combined remedy. In the future, dredging will continue to play an important role in the management of Superfund megasites and should be viewed as one of several approaches that may be necessary for their cleanup.

CONCLUSIONS AND RECOMMENDATIONS

The committee envisions that some combination of dredging, capping or covering, and natural recovery will be involved at all megasites. In situ treatments may also be required at many sites. Thus, all remedial approaches should be considered in the site evaluation, and the interactions among the various approaches should be well understood. Dredging for mass removal itself may be attractive from the viewpoint of the public, but it alone does not necessarily produce risk reduction. A better appreciation of the existing risks before dredging and what is required to achieve desired risk reduction, both in the short term and in the long term, is needed.

The challenge to this committee was twofold: to make pertinent technical recommendations (contained in Chapters 4 and 5) and to recommend changes in the management of the Superfund program that will ensure that the technical recommendations are implemented. The committee believes that three kinds of changes are critical to improve decision-making and increase dredging-remedy effectiveness at contaminated sediment megasites.

First, owing to the complexity, large spatial scale, and long time frame involved, the management of contaminated megasites should embrace a more flexible and adaptive approach to accommodate unexpected conditions and events, new knowledge, technology changes, and results of field pilot tests.

Second, improved risk assessment should specifically consider the full range and real-world limitations of remedial alternatives to allow valid comparisons of technologies and uncertainties.

Third, EPA needs a centralized focal point for coordinated assessment of contaminated sediment megasites for better consistency in site evaluations, remedy selection, and for increased focus and communication among EPA management and technical staff on what works and why.

Similar recommendations have been discussed and developed by other groups[3], but it is hoped that in the aggregate the committee's rec-

---

[3]The notion that large, complex sites need a more adaptive approach to remedy selection and implementation was the topic of much discussion at the meetings of the EPA National Advisory Council for Environmental Policy and Technology Superfund Subcommittee (NACEPT 2004). The need for adaptive-management approaches at complex contaminated sites has also been discussed in the academic community for some years (for example, Cannon 2005). Various National Research Council committees and other independent reviews have advocated similar approaches. For example, the National Research Council advocated the use of flexible phased implementation and adaptive management in environmental remediation (NRC 2001, 2003, 2005), recommended that the wide array of risks associated with implementing a remedy be explicitly considered (NRC 2001, 2005), and recommended that the limitations associated with dredging and the potential for production of residual contaminated sediment be con-

ommendations add more specificity than past efforts regarding the effective remediation and management of contaminated sediment. The three recommendations, which are described in more detail below, will in some cases require additional resources and, equally challenging in large organizations, new ways of doing business. The committee cannot stress enough that because of the potentially huge cost and the complexity of sediment megasites, the costs and efforts required to change standard operating procedures are worth the up-front investment that will be required. As noted, many times cleaning up sediment megasites sites may take decades from investigation to cleanup and cost hundreds of millions of dollars. The cost of implementing the recommendations in this report should be viewed in that context. In fact, the committee is concerned that if its recommendations are not implemented, many hundreds of millions of dollars of government and private funds will be wasted on ineffective remedies for contaminated sediment megasites.

1.   An adaptive management approach is essential to the selection and implementation of remedies at contaminated sediment megasites where there is a high degree of uncertainty about the effectiveness of dredging.

If there is one fact on which all would agree, it is that the selection and implementation of remedies at contaminated sediment sites are complicated. Many large and complex contaminated sediment sites will take years or even decades to remediate and the technical challenges and uncertainties of remediating aquatic environments are a major obstacle to cost-effective cleanup.

Because of site-specific conditions—including hydrodynamic setting, bathymetry, bottom structure, distribution of contaminant concentrations and types, geographic scale, and remediation time frames—the remediation of contaminated sediment is neither simple nor quick, and the notion of a straightforward "remedial pipeline" that is typically used to describe the decision-making process for Superfund sites is likely to be at best not useful and at worst counterproductive.

---

sidered (NRC 1997, 2001). Yet, at the time of the present review, little progress has been made in implementing those recommendations.

The typical Superfund remedy-selection approach, in which site studies in the remedial investigation and feasibility study establish a single path to remediation in the record of decision, is not the best approach to remedy selection and implementation at these sites owing to the inherent uncertainties in remedy effectiveness. At the largest sites, the time frames and scales are in many ways unprecedented. Given that remedies are estimated to take years or decades to implement and even longer to achieve cleanup goals, there is the potential—indeed almost a certainty—that there will be a need for changes, whether in response to new knowledge about site conditions, to changes in site conditions from extreme storms or flooding, or to advances in technology (such as improved dredge or cap design or in situ treatments). Regulators and others will need to adapt continually to evolving conditions and environmental responses that cannot be foreseen.

These possibilities reiterate the importance of phased, adaptive approaches for sediment management at megasites. As described previously, adaptive management does not postpone action, but rather supports action in the face of limited scientific knowledge and the complexities and unpredictable behavior of large ecosystems [NRC, 2004].

2. EPA should compare the net risk reduction associated with the various remedial alternatives, taking into account the limitations of each approach in selecting site remedies, such as residuals and resuspension.

One subject of great interest and concern at contaminated sediment Superfund sites is the risk-based comparison of remedial alternatives to support selection of a remedy (Bridges et al. 2006; Wenning et al. 2006). The committee was charged only with evaluating the effectiveness of dredging and not with comparing the effectiveness of remedial alternatives. However, the committee recognizes that the effectiveness of a dredging remedy depends on good planning, and good planning includes an evaluation of net risk reduction associated with each remedial alternative. Therefore, the committee recommends evaluating the net risk reduction of remedial alternatives to facilitate scientifically based decision making at megasites.

Baseline risk is quantified in the remedial investigation for all NPL sites, but the feasibility study may or may not include a quantitative estimate of the risks posed by alternative remedies. EPA (2005, p. 2-14) indicates that "although significant attention has been paid to evaluating baseline risks, traditionally less emphasis has been placed on evaluating risks from remedial alternatives, in part because these risks may be difficult to quantify." Even if such quantitative comparative risk assessment is provided, it might be limited in scope. For example, the feasibility study for the Upper Hudson River (TAMS Consultants 2000) included a quantitative comparison of human health and ecologic risk reduction for the fish-consumption exposure pathway, the pathway associated with the highest risk estimates. However, the analysis did not quantify short-term effects on the local community or workers or other effects that might occur during dredging; it concluded that "there is no reliable means of quantifying potential short-term impacts from activities such as sediment resuspension, habitat loss, or other transient effects."

Each remedial alternative offers its own set of risk-reduction benefits and possibly the creation of new exposure pathways and associated risks. A confounding issue is that site conditions can change in ways that help to reduce risk. That would be the case, for example, with deposition of cleaner material over residual contamination. Site-specific measurements and models need to incorporate an understanding of such site features both spatially and temporally to support valid comparisons of remedial alternatives.

Environmental responses to remediation, including sediment and biota concentration changes, are complex and difficult to predict. During remedy selection, the uncertainty around estimates of responses to remediation should be recognized and quantified to the extent warranted to optimize decision-making. For example, EPA established a tiered approach to probabilistic risk assessment in the Superfund program, as shown in Figure 6-1 (EPA, 2001). Using that approach, one proceeds from a less expensive point estimate sensitivity analysis to more expensive and time-consuming quantitative uncertainty-analysis methods. The question is: When are the more advanced methods useful or necessary? Box 6-3 illustrates a situation in which additional quantitative analysis might be warranted. It presents an idealized comparison of risk estimates, including inherent uncertainty, for two remedial alternatives.

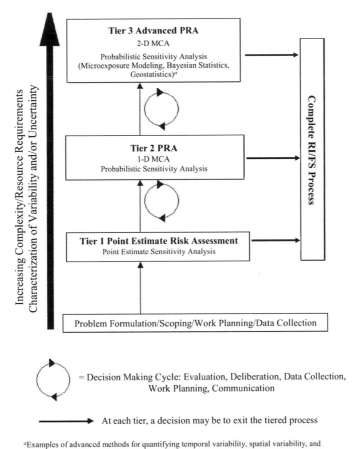

FIGURE 6-1 EPA's tiered approach to the use of probabilistic risk assessment (PRA). DMCA: decision-making cycle analysis. Source: EPA 2001.

Some potential effects will remain difficult to accurately quantify and compare (for example, the impact of a large dredging project on quality of life issues such as noise or light pollution) or potential psychological consequences from not implementing a removal remedy (for example, if community members perceive that an unmitigated threat to human health exists in their environment). Other "implementation risks" (risks potentially imposed by the implementation of a remediation strategy) such as worker and community health and safety, equipment failures, and accident rates associated with an active remediation are given little consideration in EPA's feasibility studies at Superfund sites (Wenning et al. 2006). Cura et al. (2004) identify several challenges associated with comparative risk assessment, given data limitations and the unavoidably subjective nature of quantifying some risks associated with dredged-material management decisions. However, ignoring those types of risk in comparisons of remedial options is not the solution and may have undesirable consequences, particularly when the cost of being wrong is high (Bridges et al. 2006).

BOX 6-3 Importance of Quantifying Uncertainty in Risk Estimates

In the hypothetical case outlined in the figure below, remedial alternative 1 appears to result in lower risk than alternative 2. If the uncertainty in these risk estimates is not quantified in some way, a risk manager might proceed with alternative 1. However, the uncertainty in this estimate is sufficiently high that one cannot be certain about this conclusion, and in fact a higher risk might result from implementing alternative 1. Such an outcome is obviously not desirable and even more problematic if alternative 1 is the more costly remedial alternative. With the benefit of the quantitative uncertainty analysis, and depending on the magnitudes of risk estimates and remedy costs, a risk manager might elect to gather more data (for example, with a pilot field test) to reduce uncertainty in the risk estimate for remedial alternative 1 before making a selection. A quantitative uncertainty analysis can reveal significant contributors to uncertainty in risk estimates, which are the most useful subjects of further study and data collection.

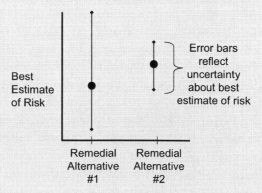

Hypothetical comparison of risk predicted for two remedial alternatives, including quantification of uncertainty associated with the risk estimates.

3. There is a great need for centralized EPA resources, responsibility, and authority at the national level to ensure that necessary improvements are made so that contaminated sediment megasites are remediated as effectively as possible.

As discussed in Chapter 2 and elsewhere in this report, it became abundantly clear during the committee's work that EPA has not devoted adequate resources and senior management attention to the issue of contaminated sediment, given the scope of the problem and the huge costs incurred by the federal government, the private sector, and others. If the recommendations in this report (and the many good reports that have gone before) are to be successfully implemented, some group in the Superfund program should be given the resources and responsibility to make needed changes and should then be held accountable.

EPA is in the best position to gather and evaluate relevant data on a national level, so it is natural for EPA to lead the effort in monitoring the progress and sharing experiences on dredging at megasites. Because every EPA region has on-the-ground experience with dredging at some megasites, regular review and shared experience can inform decision-making and raise the overall level of technical expertise. Whether by a more robust Contaminated Sediments Technical Advisory Group or some other mechanism, a consistent set of design and monitoring principles should emerge and grow from such efforts. Such information should be publicly available. The goal is generating a greater understanding of sound remediation principles and best practices and their uniform application among sites. The difficulty that the committee had in obtaining information about Superfund contaminated sediment sites and the lack of consistent data on those sites point to a need for a much stronger national program that has the authority and responsibility for overseeing and evaluating EPA's Superfund contaminated sediment efforts. The recommendations made here and by many earlier independent evaluations are unlikely to be implemented by the current patchwork approach to managing contaminated sediment sites. Resources, authority, and strong leadership are needed to ensure that the recommendations in this and prior reports are implemented in a timely manner.

It is impossible to identify a focal point for contaminated sediment sites in the current Superfund office organizational structure. Yet those sites are among the most challenging and expensive sites on the NPL.

Years of experience suggests that to garner the needed resources, focus, and management attention for a problem of this magnitude, it is necessary not only to create a "critical mass" of personnel and expertise but to clearly identify those responsible and accountable for implementing needed changes and policies. The committee strongly recommends that this gap be addressed.

Specific responsibilities include the following:

- Gather data to define the scope of the contaminated sediment problem.
- Track current and likely future contaminated sediment megasites that are on the NPL and in other EPA programs.
- Review site studies, remedies, and monitoring approaches at contaminated sediment megasites to assess whether best practices are being implemented, including whether regions are complying with national sediment and other program guidance.
- Ensure that adaptive-management approaches are applied at contaminated sediment megasites where there is substantial uncertainty about the effectiveness of dredging and other remedial approaches. As part of this effort, it is critical that EPA staff communicate clearly to local citizens and other stakeholders objective information about what dredging and other remedial options can and can not accomplish, as well as inform them about the inherent uncertainties of remedy effectiveness at sediment sites. For an adaptive-management approach to be successful, public involvement should occur "early and often."
- Ensure adequate pre- and post-remediation monitoring at complex contaminated sediment sites.
- Evaluate the effectiveness of sediment remediation in near "real time" at major sediment cleanup projects to determine whether selected remedies are achieving their intended goals and to develop lessons learned.
- Create a centralized, easily accessible, and up-to-date repository of relevant data and lessons learned regarding sediment remedies, including dredging and other approaches to facilitate information transfer among regional and headquarters staff working on these sites and the public.
- Develop and implement a research strategy for evaluating ways to improve the assessment, monitoring, and cleanup of contaminated

sediment sites, including the development and testing of new technologies.

• Serve as a focal point for coordination and communication among the many EPA programs and federal agencies who are involved in the cleanup of contaminated sediment sites.

For these functions to be implemented, EPA Headquarters program staff will need to review and provide input to regional decisions, and present senior program managers with data and information about remedy effectiveness and the approaches tried at different sites on an ongoing (and real time) basis. This will require the commitment of senior mangers in EPA Headquarters and the 10 EPA regional offices to work together.

Some of the specific tasks that will be needed to implement the above functions are described in more detail below.

EPA should define the scope of the problem.

One of the necessary tasks will be to define the scale of the megasite-sediment problem. As noted in Chapter 1 of this report, EPA has attempted to define the extent of contaminated sediment in the United States since at least the 1970s. The latest report (EPA 2004), based on the National Sediment Inventory (NSI) database, surveys about 9% of the water-body segments in the United States and classifies 43% of this nonrandom sample as having probable associated adverse effects. However, EPA's efforts have fallen short of the systematic assessment needed to define the scope of contaminated sediment that may require remedial action.[4] Even at Superfund sites, EPA's efforts to determine the geographic extent and volume of contaminated sediment appears episodic,

---

[4]Similarly, a recent EPA Office of the Inspector General report (EPA 2006, P.19) concluded that "EPA's 2004 National Sediment Quality Survey report did not provide a complete assessment of the extent and severity of sediment contamination across the Nation, nor did it fully meet the requirements of the Water Resources Development Act. . . . As a result, EPA cannot accurately estimate the volume and risks posed by contaminated sediments on a national scale. Such a national assessment would better enable EPA to ensure that it devotes resources to contaminated sediment issues that pose the greatest risks to human health and the environment."

and, as described in Chapter 2, there is no current list of contaminated sediment sites, nor does the Agency evaluate new NPL sites when they are listed to develop a "watch list" of those sites that are likely to be future megasites. In this regard, conclusions of one of EPA's earliest reports on the subject (EPA 1987) still holds true: "Although it is reasonable to say that there is significant in-place contamination in U.S. waters, it is not possible with the current level of knowledge to quantify the problem. We do not know and cannot even begin to estimate, for example, the river miles affected or the cubic yards of sediment involved." Of course EPA can not and should not wait until it has compiled a definitive picture of the contaminated sediments problem in the United States to move forward with site cleanups.

Defining the scope of the sediment problem is important for two reasons. First, it will help to place the magnitude of the problem in proper perspective to help in understanding how much of the problem has already been addressed and how much remains to be done. A concrete goal for EPA should be to have an on-going process of evaluating newly listed NPL sites, as well as major contaminated sediment sites addressed by other programs, to understand the magnitude and severity of contaminated sediments. From those evaluations, the agency should produce a report that describes the number of past, active, and probable future contaminated sediment Superfund sites and the number of likely megasites. The report should describe the types of contaminants and the volumes of contaminated sediment and lay out an estimate of likely future costs of cleanup and long-term monitoring. This kind of information was not available from EPA, which made it difficult to understand the scope of the problem.

Second, documenting how much work remains to be done and at what cost should help senior EPA management and other officials to identify the most pressing program and research needs. For example, if many more site remedies remain to be executed or listed on the NPL, it makes sense to invest in developing new technologies for remediation. If few remedies remain to be chosen, then it may be that developing monitoring tools is of greater importance.

All the recommendations in this report will take staff time and money to implement. By clearly defining the scope of the problem, EPA management will have the information it needs to identify the most important tasks to accomplish the goal of improving the scientific basis of selecting the most effective remedies for contaminated sediment sites.

EPA should ensure that adequate monitoring strategies are implemented.

As highlighted in Chapters 4 and 5, one difficulty in understanding the effectiveness of dredging is that statistically valid baseline pre-dredging condition assessments are generally not done. Without adequate pre-dredging and post-dredging monitoring, it is impossible to make the valid comparisons that are necessary to support definitive statements about the degree to which remedial objectives have been attained as a result of dredging. Much greater attention should be given to sufficient monitoring to allow valid statistical comparisons of conditions before and after dredging. That will require considerable forethought in sampling design, sampling methods, and analytic techniques. The long period from site investigation through remedial action to required 5-year reviews compounds the problem.

All that points to the need for EPA (and hopefully other federal agencies with a stake in this arena, for example, the Army Corps and the Navy) to invest in better and more consistent measuring tools to monitor conditions in the field more reliably and efficiently. The committee recommends greater efforts to develop better methods to measure sediment stability and transport processes, biogeochemical processes, and pore water concentrations and fluxes.

EPA should develop and implement a contaminated sediment research and evaluation strategy.

One of the key elements of an improved sediment-cleanup program is to establish a coherent research and evaluation strategy to fill critical information gaps. In Chapter 4, the committee reached some specific conclusions regarding factors that contribute to or limit dredging effectiveness. The EPA research and evaluation strategy should build on the work of the committee to ensure that experience gained in dredging in a variety of combinations and situations is translated into useful guidance to EPA regions and communicated to the full panoply of external stakeholders in a timely and transparent fashion.

The objective of the research and evaluation program should be to answer key questions as to what risk reduction will be achieved by different technical approaches, under what site conditions, and with what certainty. The agency needs to answer those questions through pilot studies and data collection efforts that monitor baseline to long-term conditions and that stress robust sampling and statistical analysis. To this end, EPA should undertake or commission real-time independent evaluations of the effectiveness of dredging and other remedies at contaminated sediment sites, especially megasites. The reviews would build on the committee's analyses and assess the effectiveness of dredging and other remedies at all major sediment cleanup projects and seek to understand the factors that contributed to or limited the effectiveness of the cleanup approach. This kind of study should either be conducted by a neutral external organization (either academic or non-profit) or, if conducted by EPA, be made subject to external peer review. It should be clear at the outset that an external organization conducting the review will have full control of the results, and that the final report will be made publicly available.

This type of systematic evaluation will require EPA's Superfund office and Office of Research and Development to work together to fill the critical gaps in guidance and standard protocols. This effort should also involve other agencies, such as the U.S. Geological Survey, the Army Corps of Engineers, the U.S. Navy, and the National Oceanic and Atmospheric Administration, who work on contaminated sediment sites. To implement a unified research and evaluation strategy successfully, EPA will need to ensure that appropriate resources are applied and that the various EPA and other government offices involved are held accountable for timely implementation of the strategy once it has been developed.

While there are only a few general approaches to sediment remediation, there is room for improvement in their performance. Improving and optimizing remediation systems has long been a cornerstone in environmental engineering and remediation. The refinement, modification, and development of sediment remediation approaches and technologies can overcome limitations to remedial performance and improve effectiveness. Therefore, research to improve and develop new remediation technologies, site-characterization techniques, and monitoring tools is essential to advance sediment remediation and should be supported.

Efforts to understand and promote those practices and operations that improve remediation effectiveness, and the training of decision makers and practitioners in those operations, is also critical to advance the field and improve the performance of remedial operations.

In sum, EPA's efforts should focus on moving forward with remedies at sites and, at the same time on investing the effort needed to make sure that each new pilot test or remedy implemented increases our collective knowledge of what works and what does not work and why. Because many of contaminated sediment sites are vast and remediating them will be expensive, it is worth investing time and resources now to try to ensure more cost-effective remedies in the future. Such a focus is needed if the country is to make the best possible use of the billions of dollars that will be spent on site remediation.

REFERENCES

Alcoa Inc. 2005. Remedial Options Pilot Study Work Plan: Grasse River Study Area Massena, New York. Alcoa Inc., Massena, New York.. February 11, 2005.

Bridges, T.S., S.E. Apitz, L. Evison, K. Keckler, M. Logan, S. Nadeau, and R.J. Wenning. 2006. Risk-based decision making to manage contaminated sediments. Integr. Environ. Assess. Manage. 2(1):51-58.

Cannon, J.Z. 2005. Adaptive management in superfund: Thinking like a contaminated site. New York U. Environ. Law J. 13(3):561-612.

Cura, J.J., T.S. Bridges, and M.E. McArdle. 2004. Comparative risk assessment methods and their applicability to dredged material management decision-making. Hum. Ecol. Risk Assess. 10(3):485-503.

Dickerson, D., and J. Brown. 2006. Case Study on New Bedford Harbor. Presentation at the First Meeting on Sediment Dredging at Superfund Megasites, March 22, 2006, Washington DC.

EPA (U.S. Environmental Protection Agency). 1987. An Overview of Sediment Quality in the United States. EPA-905/9-88-002. Office of Water Regulations and Standards, U.S. Environmental Protection Agency, Washington, DC, and Region 5, Chicago, IL. June 1987.

EPA (U.S. Environmental Protection Agency). 2001. Risk Assessment Guidance for Superfund: Vol. III - Part A, Process for Conducting Probabilistic Risk Assessment (RAGS 3A). EPA 540-R-02-002. Office of Emergency and Remedial Response, U.S. Environmental Protection Agency, Washington, DC. December 2001 [online]. Available: http://www.epa.gov/oswer/riskassessment/rags3adt/index.htm [accessed May 14, 2007].

EPA (U.S. Environmental Protection Agency). 2004. The Incidence and Severity of Sediment Contamination in Surface Waters of the United States: National Sediment Quality Survey, 2nd Ed. EPA-823-R-04-007. Office of Science and Technology, U.S. Environmental Protection Agency, Washington, DC. November 2004 [online]. Available: http://www.epa.gov/water science/cs/report/2004/nsqs2ed-complete.pdf [accessed Dec. 26, 2006].

EPA (U.S. Environmental Protection Agency). 2005. Contaminated Sediment Remediation Guidance for Hazardous Waste Sites. EPA-540-R-05-012. OSWER 9355.0-85. Office of Solid Waste and Emergency Response, U.S. Environmental Protection Agency. December 2005 [online]. Available: http://www.epa.gov/superfund/resources/sediment/pdfs/guidance.pdf [accessed Dec. 26, 2006].

EPA (U. S. Environmental Protection Agency). 2006. EPA Can Better Implement Its Strategy for Managing Contaminated Sediments. Evaluation Report No. 2006-P-00016. Office of Inspector General, U. S. Environmental Protection Agency, Washington, DC. March 15, 2006 [online]. Available: http://www.epa.gov/oig/reports/2006/20060315-2006-P-00016.pdf [accessed Oct. 22, 2007].

Foth & Van Dyke. 2000. Summary Report: Fox River Deposit N. Wisconsin Department of Administration, Wisconsin Department of Natural Resources. April 2000 [online]. Available: http://dnr.wi.gov/org/water/wm/foxriver/documents/sediment/depositn_report.pdf [accessed Jan. 10, 2007].

Linkov, I., D. Burmistrov, J. Cura, and T.S. Bridges. 2002. Risk-based management of contaminated sediments: Consideration of spatial and temporal patterns in exposure modeling. Environ. Sci. Technol. 36(2):238-246.

Linkov, I., F.K. Satterstrom, G. Kiker, C. Batchelor, T. Bridges, and E. Ferguson. 2006a. From comparative risk assessment to multi-criteria decision analysis and adaptive management: Recent developments and applications. Environ. Int. 32(8):1072-1093.

Linkov, I., F.K. Satterstrom, G.A. Kiker, T.S. Bridges, S.L. Benjamin, and D.A. Belluck. 2006b. From optimization to adaptation: Shifting paradigms in environmental management and their application to remedial decisions. Integr. Environ. Assess. Manage. 2(1):92-98.

Montgomery Watson. 2001. Final Summary Report: Sediment Management Unit 56/57 Demonstration Project, Fox River, Green Bay, Wisconsin. Prepared for Fox River Group and Wisconsin Department of Natural Resources, By Montgomery Watson. September 2001 [online]. Available: http://dnr.wi.gov/org/water/wm/foxriver/documents/finalreport/final_summary_report.pdf [accessed Jan. 10, 2007].

NACEPT (National Advisory Council for Environmental Policy and Technology). 2004. Final Report. Superfund Subcommittee of National Advisory Council for Environmental Policy and Technology. April 12, 2004. 264pp [online]. Available: http://www.epa.gov/oswer/docs/naceptdocs/NACEPT superfund-Final-Report.pdf [accessed May 14, 2007].

NRC (National Research Council). 1997. Contaminated Sediments in Ports and Waterways. Washington, DC: National Academy Press.

NRC (National Research Council). 2001. A Risk-Management Strategy for PCB-Contaminated Sediments. Washington, DC: National Academy Press.

NRC (National Research Council). 2003. Environmental Cleanup at Navy Facilities: Adaptive Site Management. Washington DC: National Academies Press.

NRC (National Research Council). 2004. Adaptive Management for Water Resources Project Planning. Washington, DC: National Academies Press.

NRC (National Research Council). 2005. Superfund and Mining Megasites – Lessons from the Coeur d'Alene River Basin. Washington DC: National Academies Press.

TAMS Consultants, Inc. 2000. Hudson River PCBs Reassessment RI/FS Phase 3 Report: Feasibility Study. Prepared for U.S. Environmental Protection Agency, Region 2 and U.S. Army Corps of Engineers, Kansas City District. December 2000 [online]. Available http://www.epa.gov/hudson/fs000001. pdf [accessed May 14, 2007].

von Stackelberg, K.E., D. Burmistrov, D.J. Vorhees, T.S. Bridges, and I. Linkov. 2002. Importance of uncertainty and variability to predicted risks from trophic transfer of PCBs in dredged sediments. Risk Anal. 22(3):499-512.

Wenning, R.J., M. Sorensen, and V.S. Magar. 2006. Importance of implementation and residual risk analyses in sediment remediation. Integr. Environ. Assess. Manag. 2(1):59-65.

WI DNR/EPA (Wisconsin Department of Natural Resources and the U.S. Environmental Protection Agency). 2002. Record of Decision, Operable Unit 1 and Operable Unit 2, Lower Fox River and Green Bay. Wisconsin Department of Natural Resources, Madison, WI, and the U.S. Environmental Protection Agency, Region 5, Chicago, IL. December 2002 [online]. Available: http://dnr.wi.gov/org/water/wm/foxriver/documents/whitepapers/Final_R ecord_of_Decision_OUs_1-2.pdf [accessed Jan. 11, 2007].

Appendix A

Statement of Task for the Committee on Sediment Dredging at Superfund Megasites

An NRC committee will conduct an independent evaluation of dredging projects that will look at the expected effectiveness of dredging contaminated sediments at Superfund megasites. The assessment will consider whether EPA's estimated risk reduction benefits are likely to be achieved in the time frame as predicted. Aspects of risk reduction include decreased potential for current and long-term exposure of human and ecological receptors and decreased potential for environmental dispersion of contaminants. The assessment will also consider the potential for short-term increases in risks due to resuspension during dredging. The committee will consider sites where information is available for assessing dredging effectiveness. It will strive to develop recommendations that will facilitate scientifically based and timely decision making for megasites in the future. In doing so, the committee will consider whether current monitoring regimens are sufficient to inform assessments of effectiveness and what practices should be implemented in monitoring strategies. The committee will not recommend particular remedial strategies at specific sites. The committee's considerations will include:

• Whether planned sediment cleanup levels have been reached and maintained after dredging.

- If the predicted magnitude and timing of risk reduction as a result of dredging are likely to be achieved.

- The key site-specific factors that contribute most to achieving high dredging effectiveness.

- The short-term and long-term impacts on ecologic communities as a result of dredging.

- Monitoring strategies in use and proposed for use at dredging sites and whether these strategies are sufficient to inform assessments of effectiveness.

- The specific types of assessments useful for measuring effectiveness, in particular, measuring the reduction of risk.

- How conclusions about completed and ongoing dredging operations can inform decisionmaking in the future.

It is expected that sources of information available for this assessment would include megasites for which dredging has been completed; megasites for which plans have been developed; partially implemented, and operations are ongoing; and smaller sites that exhibit lessons relevant to megasites.

Appendix B

Biographic Information on the Committee on Sediment Dredging at Superfund Megasites

Charles O'Melia (Chair) is the Abel Wolman Professor of Environmental Engineering and chair of the Department of Geography and Environmental Engineering at Johns Hopkins University, where he has served on the faculty for over 25 years. Dr. O'Melia's research fields include aquatic chemistry, environmental colloid chemistry, water and wastewater treatment, modeling of natural surface and subsurface waters, and the behavior of colloidal particles. He has served on the advisory board and review committees for the environmental engineering departments of multiple universities. He has advised professional societies, including the American Water Works Association and Research Foundation, the Water Pollution Control Federation, the American Chemical Society, and the International Water Supply Association. He served the U.S. Environmental Protection Agency as a peer-review panel member and on the Science Advisory Board. Dr. O'Melia has consulted for a variety of municipal, industrial, and government clients. In addition, he has served on several National Research Council committees, including chairing the Steering Committee for the Symposium on Science and Regulation and the Committee on Watershed Management for New York City. He was also a member of the National Research Council Water Science and

Technology Board and Board on Environmental Studies and Toxicology. Dr. O'Melia earned a PhD in sanitary engineering from the University of Michigan. In 1989, Dr. O'Melia was elected to the National Academy of Engineering for significant contributions to the theories of coagulation, flocculation, and filtration leading to improved water-treatment practices throughout the world.

G. Allen Burton is a professor of environmental sciences and director of the Institute for Environmental Quality at Wright State University. He has served as a NATO senior research fellow in Portugal and a visiting senior scientist in Italy and New Zealand. He was the Brage Golding Distinguished Professor of Research at Wright State University. Dr. Burton's research during the last 25 years has focused on developing effective methods for identifying ecologic effects and stressors in aquatic systems where sediment and storm-water contamination is a concern. His ecosystem risk assessments have evaluated multiple levels of biologic organization, from microbial to amphibian effects. Dr. Burton serves on numerous national and international scientific committees, review panels, councils, and editorial boards, and he consults for industry and regulatory agencies. He earned his PhD in environmental science (aquatic toxicology) from the University of Texas at Dallas.

William Clements is a professor at Colorado State University, where he has served on the faculty since 1989. Dr. Clements's primary research interests are in basic aquatic ecology and ecotoxicology. His research has focused on understanding how benthic macroinvertebrate communities respond to natural and anthropogenic stressors. More recently, his research projects have included assessments of recovery from fire disturbance, quantifying interactions between natural and anthropogenic stressors, and measuring abiotic factors that influence contaminant bioavailability. Dr. Clements has a substantial record of publication on benthic invertebrates, benthic community interactions, and effects of stressors. He is the author of several book chapters and a coauthor of the book *Community Ecotoxicology,* published in 2002. Dr. Clements earned his PhD in zoology from Virginia Polytechnic Institute and State University in 1988.

Frank C. Curriero in an assistant professor in the Departments of Environmental Health Sciences and Biostatistics at the Bloomberg School of Public Health, Johns Hopkins University. His research expertise and interests include applications of spatial statistics and geographic information systems for environmental public health. Dr. Curriero's research has spanned applications involving environmental epidemiology, disease mapping, spatial variation in risk and exposure assessment models, and geostatistical methods. His current methodologic research includes statistical methods for censored spatial data and models for non-Euclidean isotropic spatial dependence in geostatistics. Dr. Curriero earned his PhD in statistics from Kansas State University.

Dominic Di Toro is the Edward C. Davis Professor of Civil and Environmental Engineering in the Department of Civil and Environmental Engineering at the University of Delaware and is a consultant for HydroQual, Inc. Dr. Di Toro has specialized in the development and application of mathematical and statistical models to stream, lake, estuarine, and coastal water-quality and sediment-quality problems. He has participated as an expert consultant, principal investigator, and project manager on numerous water- quality studies for industry, research foundations, and government agencies. Recently, his work has focused on the development of water-quality and sediment-quality criteria, sediment-flux models for nutrients and metals, and integrated hydrodynamic, sediment-transport, and water-quality models. Dr. Di Toro received his PhD in civil and geological engineering from Princeton University. In 2005, Dr. Di Toro was elected to the National Academy of Engineering for leadership in the development and application of mathematical models for establishing water-quality criteria and making management decisions.

Norman Francingues retired from the U.S. Army Corps of Engineers in 2002 with over 30 years of federal civil service. He is the recipient of the Army Engineer Association Bronze Order of the de Fluery Medal and the Army Meritorious Civilian Service Award from the chief of engineers. He is a senior consultant with OA Systems Corporation. Mr. Francingues worked for the Army Corps as a senior technical adviser and for other national and international agencies on the environmental engineering aspects of navigation and hazardous-waste projects. He was technical lead for the development of innovative dredging technologies for the Army Corps of Engineers Dredging Operations and Environmental Research (DOER) program. He advises on contaminated dredged material for the International Navigation Association (PIANC), headquartered in Brussels, Belgium. His research involves innovative dredging technologies, fluidized-sediment evaluations, confined placement of contaminated dredged material, and treatment of contaminated sediments and soils. Mr. Francingues earned an MS in environmental engineering from Mississippi State University.

Richard Luthy is the Silas H. Palmer Professor and chair of the Department of Civil and Environmental Engineering at Stanford University. His research interests include environmental engineering and water quality, particularly phase partitioning and the treatment and fate of hydrophobic organic compounds. His research emphasizes interdisciplinary approaches to the behavior and availability of organic contaminants and the application of these approaches to bioavailability and environmental-quality criteria and sediment restoration. He chaired the National Research Council's Water Science and Technology Board and its Committee on Bioavailability of Contaminants in Soils and Sediments. He is a past president of the Association of Environmental Engineering Professors. He is a registered professional engineer and a Diplomate of the American Academy of Environmental Engineers. He received his PhD in environmental engineering from the University of California, Berkeley. Dr. Luthy was elected a member of the National Academy of Engineering in 1999 for leadership in the treatment of industrial wastewaters, contaminated soils, and aquifers.

Perry L. McCarty is the Silas H. Palmer Professor Emeritus in the Department of Civil and Environmental Engineering at Stanford University. He directed the Western Region Hazardous Substance Research Center from 1989 to 2002. Dr. McCarty specializes in environmental engineering with emphasis on biologic processes for water-quality control and the control of hazardous substances in treatment systems and groundwater. His research interests over the last 45 years have been in biologic processes for the control of environmental contaminants. His early research was on anaerobic treatment processes, biologic processes for nitrogen removal, and biologic degradation of hazardous chemicals.

His current interests are in aerobic and anaerobic biologic processes for control of chlorinated solvents, advanced wastewater-treatment processes, and movement, fate, and control of groundwater contaminants. Dr. McCarty earned his ScD in sanitary engineering from the Massachusetts Institute of Technology. He was elected to the National Academy of Engineering in 1977 for contributions to the environmental engineering profession through education, research, and service to government and industry.

Nancy Musgrove is the president of Management of Environmental Resources, Inc. Ms. Musgrove is experienced as an aquatic ecologist; working with both regulators and the regulated community, she has expertise in assessment of risks to aquatic communities, water-quality and sediment- quality investigations, and design of environmental monitoring programs and laboratory and field studies. She has been involved in numerous regional and national sediment investigation and cleanup projects and the peer review of decisions made at contaminated-sediment sites. Ms. Musgrove has substantial experience with the regulatory framework and technical protocols governing environmental-management decisions throughout the United States and Canada. She earned an M.S. in fisheries from the University of Washington.

Katherine N. Probst is a senior fellow at Resources for the Future. Over the last 25 years, she has conducted numerous analyses of environmental programs, focusing mainly on improving the implementation of Superfund and other hazardous-waste management programs. She was the lead author of the study *Superfund's Future: What Will it Cost?*, requested by Congress, on the estimated cost of the Superfund program to the U.S. Environmental Protection Agency (EPA). Her most recent study, *Success for Superfund*, includes recommendations for specific information that EPA should make available to the public on all Superfund sites in a site "report card." Ms. Probst also has investigated issues related to the use of institutional controls at contaminated sites, long-term stewardship, and the cleanup of sites in the nuclear-weapons complex. She was a member of EPA's Superfund National Advisory Council for Environmental Policy and Technology Subcommittee and of the EPA Science Advisory Board committee that reviewed analyses of the benefits of the

Superfund program. Ms. Probst received an MA in city and regional planning from Harvard University.

Danny Reible joined the faculty of the University of Texas at Austin College of Engineering in 2004; he holds the Bettie Margaret Smith Chair of Environmental Health Engineering. He is also director of the Hazardous Substance Research Center/South and Southwest, a consortium of Louisiana State University, Rice University, Texas A&M University, the Georgia Institute of Technology, and the University of Texas at Austin. Dr. Reible leads both fundamental and applied efforts in the assessment and management of risks associated with hazardous substances, especially as they apply to contaminated sediments. Dr. Reible has led the development of in situ sediment capping, and he has evaluated the applicability of capping technology to a wide array of contaminants and settings, including polycyclic aromatic hydrocarbons from fuels, manufactured-gas plants, and creosote- manufacturing facilities; polychlorinated biphenyls; and metals. He has consulted for both industry and regulatory groups on the applicability and design of capping for remediation at a variety of specific sites. His research has also focused on the natural attenuation of contaminants as a result of various processes in the environment. He received his PhD in chemical engineering from the California Institute of Technology. Dr. Reible was elected a member of the National Academy of Engineering in 2005 for the development of widely used methods of managing contaminated sediments.

Louis J. Thibodeaux is the Jesse Coates Professor at the Louisiana State University College of Engineering. Dr. Thibodeaux's experience and expertise are in chemical-transport processes at and across the natural media (air, water, soil, and sediments) interfaces. Specific applications have included chemical movement associated with landfill disposal, treatment, and storage of aqueous waste. He has conducted environmental research projects on chemical spills in rivers, volatiles from wastewater, nutrient cycling/modeling in lakes, and hazardous substances in contaminated bed sediment in natural aquatic systems. His current research efforts address three key aspects of the remediation chemodynamics of bed-sediment contamination: the natural recovery processes of in situ bed-sediment in the aquatic environment of rivers, lakes, and estuaries; the processes occurring with the surface soils formed from extracted (ex situ) dredged material; and the chemodynamics associated with mud clouds produced during dredging. Dr. Thibodeaux has served the National Research Council as cochair of the Steering Committee for the National Symposium on Strategies and Technologies for Cleaning up Contaminated Sediments in the Nation's Harbors and Waterways, the Committee on Risk-Based Criteria for Non-RCRA Hazardous Waste, the Committee on Contaminated Marine Sediments, and the Committee on Remedial Action Priorities for Hazardous Waste Sites. Dr. Thibodeaux earned his PhD in chemical engineering from Louisiana State University.

Donna J. Vorhees is a principal scientist with the Science Collaborative, where she provides human health risk-assessment consulting services for a variety of municipal, federal, and industrial clients. She is also an instructor at the Boston University School of Public Health, where she teaches a course in risk- assessment methods. She has extensive experience in addressing environmental questions arising from multipathway human exposure to chemicals that have been released to indoor and outdoor environments at federal and state hazardous-waste sites. Her research interests include development of probabilistic human-exposure models; field surveys to collect data needed to support risk assessment, such as samples of biota consumed by humans and interviews with anglers regarding fish-consumption practices; identification of research priorities for improving dredged-material management; and preparation of environmental-health educational materials. Dr. Vorhees conducted probabilistic analyses of multipathway exposure to PCBs in residences near the New Bedford Harbor, MA, Superfund site, to PCBs and pesticides that accumulate in fish from an offshore dredged-material disposal site, and to PCBs, dioxins, and furans that accumulate in agricultural products from the floodplain of a contaminated river. She is an active member of the Society for Risk Analysis and the International Society of Exposure Analysis. Dr. Vorhees earned her master's degree and doctorate in environmental health from the Harvard School of Public Health.

John R. Wolfe is a senior manager at Limno-Tech, Inc. He has expertise in fate and transport modeling of contaminants and environmental economics, and he manages projects in contaminated sediment, wastewater treatment and discharge permitting, combined sewer-overflow control, and groundwater protection for a variety of municipal, state, federal, and industrial clients. Dr. Wolfe holds an MSE from the University of Michigan, where has an adjunct teaching appointment in the Department of Civil and Environmental Engineering. He also holds a PhD in economics from the University of Pennsylvania and was associate professor of economics at Michigan State University. He is a licensed professional engineer and a member of the American Academy of Environmental Engineers.

Appendix C

Summary of Remedial Action Objectives, Cleanup Levels (Numerical Remedial Goals), and Their Achievement at Sediment-Dredging Sites

Note: For additional details on sites, see Table 3-1. For abbreviations, see page 294.

Site: Bayou Bonfouca, LA
Stated Remedial Action Objectives (Related to Sediment Removal): Reduce or eliminate the potential for ingestion of carcinogens in groundwater, surface soils, and shellfish. Reduce or eliminate the direct contact threat posed by bayou sediments and onsite surficial creosote waste deposits (EPA 1987).
Stated Cleanup Levels: "Contaminated sediments will be excavated either to a depth of about 6 in. into the upper cohesive layer or until PAH contamination is less than 1,300 ppm" (EPA 1987).
Dates of Remediation: 1994-1995.
Were Cleanup Levels Achieved? No chemical confirmation samples immediately after remedy. Later sampling[1] met cleanup levels.
Were Remedial Action Objectives Achieved? Partially confirmed.[2]

---

[1]1997 sampling by Louisiana (CH2M Hill 2001); 2003 sampling by EPA (EPA 2003a); and 2006 sampling by EPA after Hurricane Katrina (CH2M Hill 2006).

Comments and Lessons Learned: Advances in dredging technology highlighted ability to dredge sediment accurately. Importance and difficulty of characterizing contaminant sediment deposits accurately.[3] Importance of backfilling. Difficulty of accessing data (see Chapter 4). Lack of planned post-dredging monitoring. Less stringent cleanup level (PAHs at 1,300 ppm).

Site: Lavaca Bay, TX
Stated Remedial Action Objectives (Related to Sediment Removal): Not a CERCLA remedy. Goals of this treatability study were as follows (Alcoa 2000): develop information to support the technical and economic evaluation of potential remedial actions; evaluate the effectiveness of dredging equipment on removal of mercury impacted sediment in the study area; evaluate potential impacts of dredging on mercury mobilization and residual sediment concentrations; and, understand the impact that dredging mercury contaminated sediment may have on mercury levels in Bay biota.
Stated Cleanup Levels: None given.
Comments: Pilot study.
Dates of Remediation: 1998.
Were Cleanup Levels Achieved? Not applicable—no cleanup levels indicated.
Were Remedial Action Objectives Achieved? Pilot-study goals apparently achieved; not expected to achieve long-term risk reduction.
Comments and Lessons Learned: (1) No significant change in average surficial sediment contaminant concentrations after dredging. (2) Advantages of pilot study for describing results of large-scale dredging at this site. (3) Evaluation of residuals after each dredging pass provided useful information on generation and concentrations.

Site: Black River, OH
Stated Remedial Action Objectives (Related to Sediment Removal): Not a CERCLA remedy. The goal of the sediment remediation project was to remove PAH-contaminated sediment to eliminate liver tumors in resident brown bullhead populations (Zarull et al. 1999).

---

[2]Fish sampling was conducted in 1996 and 1997 and resulted in lifting of fish-consumption advisory in 1998 (CH2M Hill 2001), but there is no indication that shellfish sampling has been conducted.

[3]"Detailed design investigations during the summer and fall of 1988 [post-ROD, pre-dredging] showed the volume of contaminated sediments to be approximately 150,000 cy, an increase of three times that estimated in the ROD. This dramatic volume increase resulted in a cost estimate for the selected remedy rising from approximately $55 million to about $150 million" (EPA 1990b).

Stated Cleanup Levels: No chemical specific cleanup levels given. "The primary cleanup target was the removal of sediment in the area of the former USS coke plant to 'hard bottom,' or the underlaying shale bedrock. No quantitative environmental targets or end points were established, although post-dredging sampling was required to test for remaining areas of elevated PAH concentrations" (Zarull et al. 1999).

Dates of Remediation: 1989-1990.

Were Cleanup Levels Achieved? No quantitative chemical targets, but apparently met operational targets (mass removal and dredging to bedrock).

Were Remedial Action Objectives Achieved? Short-term risk increased, but long-term risk-reduction targets met.

Comments and Lessons Learned: Need for monitoring of biota. Dredging effective, although uncertain improvement over natural attenuation. Increase in fish tumors after implementation.

Site: Outboard Marine Corporation (OMC)—Waukegan Harbor, IL

Stated Remedial Action Objectives (Related to Sediment Removal): None given.

Stated Cleanup Levels: "Sediments in excess of 50 ppm PCB will be removed from the harbor by hydraulic dredging" (EPA 1984).

Dates of Remediation: 1991-1992.

Were Cleanup Levels Achieved? No chemical confirmation samples immediately after remediation. Remedy based on assumption that removal of fine-grained "muck" overlying glacial till would achieve cleanup levels. Later sampling[4] met cleanup levels; current state unclear.[5]

Were Remedial Action Objectives Achieved? Remedial action objectives not defined. Fish-tissue concentration trends inconclusive (see text).

Comments and Lessons Learned: Insufficient pre-dredging and during-dredging data on fish concentrations to make comparison with post-dredging data. Less stringent remedial-action trigger (sediments greater than 50 mg/kg removed to depth of clean underlying geologic stratum) than for other PCB cleanups. No chemical verification samples taken on sediment immediately after dredging.

---

[4]Apparently, sampling in 1996 (EPA 1999) and in 2003 (ILDPH/ATSDR 2004) indicates PCBs in sediments at less than 50 mg/kg; however, sampling events not mentioned in 5-year reviews or site summary submitted to committee.

[5]EPA states "OMC Plant 2 is likely a continual source of PCBs to Waukegan Harbor, thus further harbor sediment sampling and analysis is likely needed to confirm whether cleanup levels are still being met" (EPA 2002).

Site: Commencement Bay—Head of Hylebos, Tacoma, WA
Stated Remedial Action Objectives (Related to Sediment Removal): Achieve "acceptable sediment quality in a reasonable time frame." *Acceptable sediment quality* is defined as "the absence of acute or chronic adverse effects on biological resources or significant human health risks." *Reasonable time frame* was further defined to be a period of 10 years to allow for natural recovery (via sedimentation) (EPA 1989c).
Stated Cleanup Levels: SQOs were established in 1989 ROD, and PCB value was modified in 1997 ESD (EPA 1997b).
Dates of Remediation: 2003-2006.
Were Cleanup Levels Achieved? Cleanup levels met in all but one area; adjacent nearshore cap was extended to address this area.
Were Remedial Action Objectives Achieved? Dredging operation recently completed; no long-term data.
Comments and Lessons Learned: Ability to meet cleanup levels under favorable conditions. Sufficient sampling and reference cores aided effective site characterization and contributed to success of dredging. Capping of one area above CULs contributed to success. Site sediment characteristics ("soft black muck" over native material) permitted overdredging and visual characterization of contaminated vs native material. Cost-plus-fee contract incentivized dredging team to implement BMPs. Pilot testing[6] indicated extent and type of debris and issues related to dredging and dredge-material handling.

Site: Commencement Bay—Sitcum Waterway, Tacoma, WA
Stated Remedial Action Objectives (Related to Sediment Removal): Achieve "acceptable sediment quality in a reasonable time frame." *Acceptable sediment quality* is defined as "the absence of acute or chronic adverse effects on biologic resources or significant human health risks." *Reasonable time frame* was further defined to be a period of 10 years to allow for natural recovery (via sedimentation) (EPA 1989c).
Stated Cleanup Levels: SQOs were established in 1989 ROD, and PCB value modified in 1997 ESD (EPA 1997b).
Dates of Remediation: 1993-1994.
Were Cleanup Levels Achieved? Cleanup levels met.[7]

---

[6]"The pilot program [260 cy removal] provided information on mechanical dredging, offloading of barges to rail cars, and rail transportation of dredged material to and into an upland landfill" (Dalton, Olmsted & Fuglevand, Inc. 2006).

[7]CULs met in all areas immediately after dredging (redredging was conducted in one subarea) exception one underpier area where natural recovery was able to achieve SQOs within allowed period (confirmed in 2003) (EPA 2006a [Commencement Bay–Sitcum Waterway, April 26, 2006]).

Were Remedial Action Objectives Achieved? Not confirmed through biologic sampling, but long-term monitoring has shown continued (10 years) compliance with effects-based and risk-based cleanup levels, which were accepted as surrogates for biologic-effects testing. No further monitoring required by EPA at the site.

Comments and Lessons Learned: Clearly distinguishable contaminated layer was valuable in achieving remedial action objectives. Combining cleanup with port redevelopment provided economies for overdredging, ensuring removal. Compliance evaluated on an area basis used probabilistic criteria (averages and upper confidence limits) that accommodated single-chemical, noncontiguous, low-level exceedances after dredging. Inclusion of natural recovery for low-level contamination in remedial options allowed reasonable response to undredged inventory in under-pier areas.

Site: Duwamish Diagonal, Seattle WA

Stated Remedial Action Objectives (Related to Sediment Removal): Not a CERCLA remedy (Natural Resource Damage Settlement). Restore and replace natural resources within the lower Duwamish River and Elliott Bay that have been injured by releases of hazardous materials through remediation of contaminated sediments in the vicinity of combined sewer overflows and storm drains, source control, and habitat restoration (U.S.A et al. v the City of Seattle, Consent Decree No. C90-395WD, December 23, 1991).

Stated Cleanup Levels: No chemical specific cleanup levels given. Dredging performance criteria based on achieved specific elevation for cap placement (EcoChem Inc. 2005).

Comments: Project viewed as source-control action to address natural resource damages at a CSO through a hot-spot cleanup, with remaining contamination addressed as part of Lower Duwamish Superfund site actions.

Dates of Remediation: 2003-2004.

Were Cleanup Levels Achieved? Dredging performed for cap placement; performance criteria based on elevation specification. Toxicity data used to define area to be dredged, although performance based on chemical criteria.

Were Remedial Action Objectives Achieved? Unlikely. Cleanup performed as interim action before selection of remedy for LDW site.

Comments and Lessons Learned: Lack of adherence to BMPs[8] resulted in significant transport of contaminated sediment outside the dredge prism. Post-dredging monitoring showed increased concentrations of PCBs and other COCs

---

[8]The sediment remediation project closure report indicated that "the most obvious problems were over-filling the dredge bucket and spilling material out of the bucket as it was moved to and from the barge." Water quality monitoring also indicated exceedances of turbidity-compliance criteria (EcoChem Inc. 2005).

in adjacent areas, which required placement of additional thin layer of clean material. Biologic monitoring conducted as part of wider LDW site indicates increases in fish-tissue contaminant concentrations at project site (see text).

Site: Puget Sound Naval Shipyard, Bremerton, WA
Stated Remedial Action Objectives (Related to Sediment Removal): Reduce risks to subsistence fishers consuming seafood from Sinclair Inlet by reducing PCB concentrations in biologically active zone of sediment in marine Operable Unit, controlling shoreline erosion of contaminated fill material, and selectively removing high concentrations of mercury that were colocated with PCBs (EPA 2000c).
Stated Cleanup Levels: PCBs at 0.023 mg/kg wet weight in fish tissue and at 3.0 mg/kg (OC normalized) in sediment on area-weighted average. Sediment remedial action objectives to be achieved within 10 years. No time frame for recovery of fish tissue (EPA 2006a [Puget Sound Naval Shipyard, May 15, 2006]).
Comments: Action levels were defined to distinguish which technology would be implemented. Dredging occurred when sediments had PCBs above 12 mg/kg (OC normalized) or above 6 mg/kg (OC) when mercury was at over 3 mg/kg. Enhanced natural recovery (thin-layer placement) was applied where PCBs were at 6-12 mg/kg OC (EPA 2006a [Puget Sound Naval Shipyard, May 15, 2006]).
Dates of Remediation: 2000-2004.
Were Cleanup Levels Achieved? No immediate post-dredging sampling. Initial long-term monitoring shows cleanup levels not met.
Were Remedial Action Objectives Achieved? Long-term monitoring shows sediment quality has not met interim target that would support achieving goals in desired 10-year period.
Comments and Lessons Learned: Lack of adherence to BMPs. TSS exceedances. Fish[9] and sediment[10] contaminant concentrations did not decrease or increased after dredging. Importance of recognizing issues with entire dredging process train beforehand.

---

[9]Fish (English sole) sampling in 2003 indicates that concentrations after dredging exceed the cleanup goal. Average PCB concentrations are similar to average pre-remediation concentrations documented by historical (1991-1997) monitoring. The average mercury concentration is slightly higher than in previous (1994) sampling (URS 2006).

[10]The post-remediation area-weighted average PCB sediment concentration of 7.4-13 mg/kg of organic carbon (OC) (90th percentile confidence interval) exceeds the pre-remediation action area-weighted average value of 7.8 mg/kg of OC calculated from data collected before remediation (URS 2006).

Site: Harbor Island—Lockheed Shipyard, Seattle, WA

Stated Remedial Action Objectives (Related to Sediment Removal): "Reduce concentrations of hazardous substances to levels which will have no adverse effect on marine organisms by eliminating the exposure pathways associated with residual concentrations of these contaminants. . . . Restore the marine habitat to its most productive condition to the extent practicable. . . . Minimize or eliminate the potential for recontamination of the cap from groundwater. . . . Achieve adequate source control to prevent recontamination" (EPA 2006a [Harbor Island Lockheed Shipyard Sediment OU, May 11, 2006]).

Stated Cleanup Levels: Arsenic at 57 mg/kg dry weight, copper at 390 mg/kg dry weight, lead at 450 mg/kg dry weight, mercury at 0.41 mg/kg dry weight, zinc at 410 mg/kg dry weight, PCBs at 12 mg/kg organic carbon normalized, LPAHs at 370 mg/kg organic carbon normalized (low-molecular-weight polynuclear aromatic hydrocarbons), HPAHs at 960 mg/kg organic carbon normalized (high-molecular-weight polynuclear aromatic hydrocarbons), and tributyltin at 76 mg/kg organic carbon normalized (EPA 1997c, 2003c, 2006a [Harbor Island Lockheed Shipyard Sediment OU, May 11, 2006]).

Comments: Cleanup levels were based on Washington State Sediment Management Standards (Apparent Effects Thresholds); TBT cleanup level was developed on basis of site-specific data for protection of invertebrates. Area background concentrations of PCBs and mercury were allowed to modify boundary (but not cleanup level within boundary) and define acceptable levels of recontamination.

Dates of Remediation: 2003-2004.

Were Cleanup Levels Achieved? Cleanup levels not met for metals, PAHs, and PCBs in some open-water areas that were to be remediated through dredging only.[11] "Enhanced natural recovery" (placement of 6 in. of sand on the sediment surface) used in some of these areas. No actions taken in two areas with single-chemical, low-level exceedances. Toe of slope at transition from dredging only to dredging and capping also did not meet CULs. This noncompliant area addressed through overplacement of cap material.

Were Remedial Action Objectives Achieved? No long-term data yet available on objective of protection and recovery of benthic community health.

Comments and Lessons Learned: Debris affected schedule and cost. Extensive sediment characterization during dredging included progress cores to assess adequacy of dredge cuts. Use of test dredge or pilot dredge would have helped

---

[11]The 2005 5-year review (EPA 2005c) stated that "a total of eight sediment samples were collected from the post-dredge surface of the channel area. . . . All analytical results were compared to the SQS [sediment quality standards] chemical criteria to evaluate compliance. . . . [F]rom eight samples, three samples exceeded the SQS for PCBs only. Three other samples . . . exceeded the SQS for a combination of COCs [chemicals of concern]."

to characterize debris. Experienced contractors successfully completed sediment handling with careful site management and successfully reduced contaminant loss. Implementation of BMPs. Technologies (WINOPS) incorporated to permit successful dredge placement. Change in dredging contracting strategy from production-based to time-and-materials-based contributed to successful remediation.

Site: Harbor Island—Todd Shipyard, Seattle, WA

Stated Remedial Action Objectives (Related to Sediment Removal): "Reduce concentrations of hazardous substances to levels which will have no adverse effect on marine organisms by eliminating the exposure pathways associated with residual concentrations of these contaminants. . . . Restore the marine habitat to its most productive condition to the extent practicable. . . . Minimize or eliminate the potential for recontamination of the cap from groundwater. . . . Achieve adequate source control to prevent recontamination" (EPA 2006a, [Harbor Island Lockheed Shipyard Sediment OU, May 11, 2006]).

Stated Cleanup Levels: Arsenic at 57 mg/kg dry weight, copper at 390 mg/kg dry weight, lead at 450 mg/kg dry weight, mercury at 0.41 mg/kg dry weight, zinc at 410 mg/kg dry weight, PCBs at 12 mg/kg organic carbon normalized, LPAHs at 370 mg/kg organic carbon normalized, HPAHs at 960 mg/kg organic carbon normalized, and tributyltin at 76 mg/kg organic carbon normalized (EPA 1997c, 2003d, 2006a [Harbor Island Lockheed Shipyard Sediment OU, May 11, 2006]).

Comments: Cleanup levels were based on Washington State Sediment Management Standards (Apparent Effects Thresholds); TBT cleanup level was developed on basis of site-specific data for protection of invertebrates. Area background concentrations of PCBs and mercury were allowed to modify boundary (but not cleanup level within boundary) and define acceptable levels of recontamination.

Dates of Remediation: 2004-2005.

Were Cleanup Levels Achieved? Mercury and PAH cleanup levels not achieved at a few locations, but concentrations were below action levels[12] and thus acceptable without additional remediation.

Were Remedial Action Objectives Achieved? No long-term data yet available on objective of protection or /recovery of benthic community health.

Comments and Lessons Learned: Extensive sediment characterization during dredging included progress cores to assess adequacy of dredge cuts. Experienced

---

[12]EPA stated that "the average (mean) concentration and the upper 95% confidence level on the mean concentration for all COCs are less than SQS chemical criteria for all analytes. Based on this statistical evaluation, Todd and EPA have concluded that the post-dredge surface in all areas of the Site meets cleanup criteria" (EPA 2006a [Harbor Island Todd Shipyards Sediment Operable Unit, May 12, 2006]).

contractors successfully completed sediment handling with careful site management to reduce contaminant loss. Implementation of BMPs throughout process train. Technologies (WINOPS) incorporated to permit successful dredge placement. Use of dredging contractor as consultant during design phase and use of environmental performance-based contracting contributed to successful remediation.

Site: Cumberland Bay, NY

Stated Remedial Action Objectives (Related to Sediment Removal): Mitigate the immediate threat to the environment posed by the PCB-contaminated sludge bed. Rapidly and significantly reduce human and environmental risks. Prevent further environmental degradation resulting from this known source of PCB contamination (NYSDEC 1997).

Stated Cleanup Levels: None given.[13]

Comments: Entire PCB-contaminated sludge bed to be removed (NYSDEC, 1997).

Dates of Remediation: 1999-2000.

Were Cleanup Levels Achieved? No cleanup levels established.

Were Remedial Action Objectives Achieved? Not determined; some residual contamination present.

Comments and Lessons Learned: Hardpan, rocks, and gulleys inaccessible to hydraulic dredge created unfavorable conditions that required multiple dredge passes and hand-held diver dredging. High residuals after initial dredging; some contamination remained at termination of project. Lack of quantitative criteria. Inadequate sampling and characterization techniques limited initial understanding of full extent of contaminated materials.

Site: Dupont—Christina River, DE

Stated Remedial Action Objectives (Related to Sediment Removal): Prevent exposure to contaminated sediments (EPA 1993a).

Stated Cleanup Levels:

Contaminant	Original site-specific Cleanup Criteria[a]	Revised Site-specific Cleanup Criteria[b]
Zinc	5,600 ppm	3,000 ppm
Lead	1,200 ppm	700 ppm
Cadmium	60 ppm	20 ppm

[a]From 1993 ROD (EPA 1993a).
[b]Original cleanup values were lowered to eliminate need for extensive long-term monitoring program that was part of 1993 ROD (EPA 2005b).

---

[13]From record of decision (NYSDEC 1997): "Question: What is the target level DEC hopes to achieve of remaining PCB contamination? Response: The NYSDEC has not set an action level. The goal of remediation is the removal of the sludge bed in its entirety."

Comments and Lessons Learned: These cleanup criteria were apparently used to delineate area for remediation. In practice, chemical analyses were not used to verify removal of contaminated sediments. Sediments were removed to the required minimum depth of 2 ft or until underlying stratum was encountered (URS 1999).

Dates of Remediation: 1999.

Were Cleanup Levels Achieved? Not determined; no confirmation samples taken after dredging and backfilling. Removal targets (elevation) were met.

Were Remedial Action Objectives Achieved? Not determined, no confirmation or long-term monitoring.

Comments and Lessons Learned: Lack of chemical confirmation sampling and long-term monitoring is problematic.[14] Dredging operation was based on removal, not on concentration. Need for source control and a reference site.

Site: Fox River (OU 1), WI

Stated Remedial Action Objectives (Related to Sediment Removal): Achieve, to the extent practicable, surface-water quality criteria throughout Lower Fox River and Green Bay. Protect humans who consume fish from exposure to COCs that exceed protective levels. Protect ecologic receptors from exposure to COCs above protective levels. Reduce transport of PCBs from Lower Fox River into Green Bay and Lake Michigan. Minimize downstream movement of PCBs during implementation of remedy (WI DNR/EPA 2002).

Stated Cleanup Levels: Dredge all sediment with PCBs at over 1 ppm or achieve a surface-weighted average concentration (SWAC) of 0.25 ppm (WI DNR/EPA 2002) (see Comments for explanation).

Comments: If after dredging is completed for OU 1, sampling shows that the 1-ppm remedial action level (RAL) has not been achieved, a SWAC of 0.25 ppm may be used to assess effectiveness of PCB removal. If that SWAC has not been achieved, the remedy provides options to reduce risk further. The first option is additional dredging to ensure that all sediments with PCBs at over 1-ppm RAL are removed throughout the particular deposit. The second option is to place a sand cover on dredged areas to reduce surficial concentrations so that a SWAC of 0.25 ppm for OU 1 is achieved (WI DNR/EPA 2002).

Dates of Remediation: 2004-present.

Were Cleanup Levels Achieved? Dredging not complete; some subunits have not achieved desired cleanup levels.

Were Remedial Action Objectives Achieved? Dredging not yet completed.

---

[14]Second 5-year review for the site indicates that vegetative cover and vegetation species composition are monitored. However, there is no systematic monitoring of chemical concentrations in sediment, water, or biota and no evaluation of toxicity end points.

Comments and Lessons Learned: Baseline monitoring, although extensive, is not sufficient to inform long-term monitoring because it began after dredging had begun at the site. Thin layer of highly contaminated sediment and residuals have limited success at reaching 1 ppm. Heterogeneity of deposits creates difficulties in defining the dredge prism. ROD permits flexibility in achieving cleanup levels and stipulates additional actions (further dredging or capping) if dredging does not achieve results.

Site: Fox River (Deposit N), WI
Stated Remedial Action Objectives (Related to Sediment Removal): Not a CERCLA remedy. Demonstration-project objectives were environmental dredging to remove contaminated sediment to specifications; protection of the river, local properties, and residents during sediment removal; safe transport and disposal of sediment; and maintenance of good local relations during the project (Foth and Van Dyke 2000).
Stated Cleanup Levels: Average residual thickness no more than 3 in. in West Lobe and no more than 6 in. in East Lobe (Foth and Van Dyke 2000).
Comments: Pilot study.
Dates of Remediation: 1998-1999.
Were Cleanup Levels Achieved? Target elevations met.[15]
Were Remedial Action Objectives Achieved? Pilot-project goals met. Not expected to achieve long-term risk reduction.
Comments and Lessons Learned: Pilot project indicated mass removal can be achieved. Release of PCBs during dredging.[16] Bedrock limited ability to dredge

---

[15]Project summary stated that "project specifications did not require either total removal of the sediment or removal to a specific PCB sediment concentration as these sediments rested on a fractured bedrock surface, preventing a dredge cut into a clean underlying layer" (Foth and Van Dyke 2000).

[16]"During dredging in 1998 (Phase I), the upstream average reported PCB water column concentration was 3.2 ng/l compared to the average downstream PCB water column concentration of 11 ng/L. The variation between the upstream PCB water column concentration and the downstream PCB water column concentration measured during dredging reflects an average increase downstream of 3.5 times the upstream value. Similar water column PCB results were obtained during Phase II and III in the 1999 dredge season. For the 1999 dredge period, the average upstream PCB water column concentration was 14 ng/L compared to the average downstream PCB water column concentration of 24 ng/L. This variation represents an increase of 1.7 times the upstream reported value. It can be concluded from this data that dredging caused an increase in PCB concentrations downstream of the dredge site" (Malcolm Pirnie, Inc. and TAMS Consultants, Inc. 2004).

completely, and residual layer was left. Post-dredging concentrations were similar to that before dredging.[17]

Site: Fox River (SMU 56/57), WI
Stated Remedial Action Objectives (Related to Sediment Removal): Not a CERCLA remedy. Demonstration-project objectives were to evaluate potential effects of large-scale dredging of PCB-contaminated sediments on the Fox River, to evaluate efficacy of large-scale dewatering and land disposal of PCB-contaminated sediments, and to evaluate potential costs of large-scale dredging, dewatering, and land disposal of PCB-contaminated sediments (Montgomery Watson 2001).
Stated Cleanup Levels: 1999 action: to depths consistent with PCBs at 1 mg/kg of sediment or less (Montgomery Watson 2001). 2000 action: total PCBs at 1 mg/kg of sediment or 10 mg/kg with at least 6 in. of clean sand backfill (Fort James Corporation et al. 2001).
Comments: Pilot study.
Dates of Remediation: 1999-2000.
Were Cleanup Levels Achieved? Cleanup levels not met in first season of dredging (1999); met in 2000 and then backfilled.
Were Remedial Action Objectives Achieved? Pilot-project goals were met; not expected to achieve long-term risk reduction.
Comments and Lessons Learned: Residual mass and PCB concentrations resulted in redredging and backfilling. Difficulties experienced in dewatering and solids handling indicated value of pilot studies and of considering full train of treatment. Dredging released PCBs to water column despite silt curtain controls. Using turbidity as an indicator of PCB transport is insufficient.[18] No sampling after sand backfilling, so final surface concentrations are not known.

Site: Ketchikan Pulp Company, Ward Cove, AK
Stated Remedial Action Objectives (Related to Sediment Removal): Reduce toxicity of surface sediments.

---

[17]"Using the 1998 data, collected just prior to dredging, the pre-dredge average PCB sediment concentration in Deposit N was 16 ppm, with a maximum concentration of 160 ppm. The post-dredge average PCB sediment concentration in Deposit N was 14 ppm, with a maximum of 130 ppm" (Foth and Van Dyke 2000).

[18]"The TSS and PCB comparison (downstream minus upstream) illustrates that TSS is not a reliable indicator of PCB transport during a dredging operation. For example, from September 1 to October 6, a period of negative TSS loading (less at the downstream than at the upstream site), the PCB loading was positive. Thus, if one is to monitor PCB transport during a remediation operation, sole reliance on turbidity or TSS measurements is inadequate" (Steur 2000).

Enhance recolonization of surface sediments to support a healthy marine benthic infaunal community with multiple taxonomic groups (EPA 2000b).
Stated Cleanup Levels: None given.
Comments: Health of benthic communities is to be assessed through toxicity testing and benthic community analyses as part of long-term monitoring.
Dates of Remediation: 2000-2001.
Were Cleanup Levels Achieved? Success of cleanup to be determined by toxicity testing and benthic community analysis. Concentration-based cleanup levels not defined.
Were Remedial Action Objectives Achieved? Monitoring is continuing; there is some initial success in reducing benthic toxicity, which was the desired objective.
Comments and Lessons Learned: Effectiveness of backfilling in reducing toxicity was demonstrated in comparison with locations without backfilling.[19] Site demonstrated use of toxicity assays and benthic community analyses as a useful indicator of ecologic improvement after remediation. Dredging and backfilling conducted in a small area compared to backfilling only and natural recovery areas.

Site: Newport Naval Complex—McCallister Point Landfill, RI
Stated Remedial Action Objectives (Related to Sediment Removal): Prevent human ingestion of shellfish impacted by sediments with COC concentrations exceeding cleanup levels. Prevent exposure of aquatic organisms to sediments with COC concentrations exceeding cleanup levels. Prevent avian-predator ingestion of shellfish impacted by sediments with COC concentrations exceeding cleanup levels. Minimize migration of sediments with COC concentrations exceeding selected PRGs to offshore areas and previously unaffected areas of Narragansett Bay. Prevent washout of landfill debris into marine environment (EPA 2000d).
Stated Cleanup Levels: Copper, 52.9 ppb in pore water; nickel, 33.7 ppb in pore water; anthracene, 513 ppb in sediment; chrysene, 1,767 ppb in sediment; fluorene, 203 ppb in sediment; total PCBs, 3,634 ppb in sediment (EPA 2000d).
Dates of Remediation: 2001.
Were Cleanup Levels Achieved? Cleanup levels confirmed analytically immediate after dredging except when bedrock encountered. Long-term monitoring

---

[19]"The 2004 [first long-term monitoring event] data indicated that conditions in the three thin-layer capping areas had generally improved, while those in three of the four natural recovery areas generally had not (the shallow natural recovery area was the exception)" (EPA 2006a, Ketchikan Pulp Company; April 26, 2006).

indicates that sediment cleanup levels have been maintained although pore water exceedances and toxicity[20] remain.

Were Remedial Action Objectives Achieved? Although narrowly defined, remedial action objectives based on exposure to sediment above cleanup levels apparently met. Long-term risk reduction inconclusive.

Comments and Lessons Learned: Cleanup levels were met although verification sampling was insufficient in some locations (hitting bedrock was considered meeting values). Pore water exceedances of cleanup levels persist after dredging although source of contamination is not clear.[21] Ability to dredge much of site from shore was advantageous. Incomplete recolonization by shellfish and submerged aquatic vegetation in near term (less than 5 years). Useful and comprehensive range of pre-monitoring and post-monitoring metrics.

Site: GM Central Foundry, St. Lawrence River, NY
Stated Remedial Action Objectives (Related to Sediment Removal): Remedial action objectives are not specifically provided. EPA does state: "Hot spots in the St. Lawrence and Raquette rivers and Turtle creek will be dredged and excavated to remove PCBs. All PCB contaminated sediments in the hot spots will be removed given the technological limitations associated with dredging" (EPA 1991).
Stated Cleanup Levels: St. Lawrence and Racquette Rivers: PCBs at 1 ppm (EPA 1991).
Dates of Remediation: 1995.
Were Cleanup Levels Achieved? Cleanup levels not met in St. Lawrence River, because of residuals, backfilling, and capping required in one area. Cleanup levels met in Racquette River.
Were Remedial Action Objectives Achieved? Dredging alone unable to achieve cleanup levels, although combination remedy effectively reduced surface concentrations.
Comments and Lessons Learned: Intensive monitoring before, during, and immediately after dredging provided useful indications of dredging effect. Sheet-pile walls limited PCB release during dredging. Inability to eliminate residuals and possible increase in residual concentration due to contaminant retention within sheet-pile walls. Lack of sediment sampling for contamination since

---

[20]"Sea urchin pore water toxicity results indicate toxicity at some stations within Groups 1, 2, 3, and 4 (TetraTech NUS, 2006)." (Groups are geographic areas with sampling stations; groups 1, 2, 3, and 4 are all the areas besides the reference area.)

[21]The first long-term monitoring report for the site (Tetra Tech NUS 2006) states that "groundwater discharge may be responsible for elevated pore water metals concentrations found at some of the near shore locations." However, the committee notes that dredging may have influenced redox conditions leading to the dissolution of metals from residual contaminated materials.

dredging in 1995 eliminates insight into concentration changes over time. No apparent trend in fish concentrations after dredging.

Site: Grasse River, NY (Non-Time-Critical Removal Action)
Stated Remedial Action Objectives (Related to Sediment Removal): Project objectives were: (1) remove the most upstream major PCB source in the Grasse River; (2) eliminate a potential source of PCB exposure to biota; reduce potential long-term risks to human health and the environment; (3) provide valuable site-specific data for use in the Analysis of Alternatives for the study area (ALCOA 1995).
Stated Cleanup Levels: None given.
Comments: Non-Time-Critical Removal Action (NTCRA).
Dates of Remediation: 1995.
Were Cleanup Levels Achieved? None provided. Average sediment concentrations decreased.[22]
Were Remedial Action Objectives Achieved? Majority of sediment removed, but corresponding reductions in fish-tissue and water-column concentrations not observed.[23]
Comments and Lessons Learned: Debris and bedrock created operational difficulties. Site served as useful pilot study; substantial useful data were collected. Pilot improved site conceptual model. Much of targeted sediment mass was removed with substantial portion of PCBs in river system.[24] Removal released PCBs to water column as evidenced by increased water concentrations and accumulation by caged fish adjacent to work zone. Turbidity release did not correlate with PCB release. Project demonstrated value of caged-fish studies.

---

[22]According to BBL (1995): "Area A [the largest area] pre-NTCRA samples contained PCB concentrations ranging from non-detect (MDL varied) to 11,000 mg/kg, while post-removal PCB samples contained PCB concentrations ranging from 1.1 to 260 mg/kg. The arithmetic (geometric) average PCB concentration detected within area A (considering all depths) was reduced from 1,109 mg/kg (263 mg/kg) to 75 mg/kg (40 mg/kg), or about 93% (85%). Considering PCBs contained within the bioavailable zone (that is, top 12 in. of sediment), average arithmetic (geometric) PCB concentrations were reduced from 518 mg/kg (195 mg/kg) to 75 mg/kg (40 mg/kg) or approximately 86% (80%). At several locations, PCB concentrations in the residual sediment (all bioavailable) increased from pre-remediation bioavailable conditions." In the smaller area B, the pre-NTCRA arithmetic (geometric) average sediment PCB concentration of 300 mg/kg (275 mg/kg) was decreased to 108 mg/kg (106 mg/kg) following dredging, approximately a 64% (62%) decrease (BBL 1995).
[23]About 84% of the original volume of sediment was removed, leaving 550 cy of the 3,500-cy sediment that was targeted. Average thickness of sediment in area was reduced from about 22 in. to 4 in. (BBL 1995).
[24]About 27% of PCB mass in Grasse River study area (BBL 1995).

Site: Grasse River, NY

Stated Remedial Action Objectives (Related to Sediment Removal): Not a CERCLA remedy. Demonstration project objectives were (1) evaluate dredging as a remedial option to reduce the potential risk that may be posed by future ice jam related sediment scour events by removing a targeted area of sediments with elevated PCB concentrations in an area of the river that is known to be subject to ice jam-related scour; (2) develop site-specific information related to dredging effectiveness, dredging residuals, dredging production rate, and sediment resuspension that can be used in the development of the revised Analysis of Alternatives Report (Alcoa Inc. 2005).

Stated Cleanup Levels: Not a CERCLA remedy. No chemical specific cleanup levels given. Remove all soft sediments, to the extent possible, from an approximate 8-acre area of the main channel. To the extent practical, all soft sediments will be removed to hard bottom leaving a stable dredge face on the adjacent sediments (Alcoa Inc. 2005).

Comments: Pilot study.

Dates of Remediation: 2005.

Were Cleanup Levels Achieved? No site-specific cleanup levels.

Were Remedial Action Objectives Achieved? Stated goals were achieved. Not expected to achieve long-term risk reduction.

Comments and Lessons Learned: Debris and bedrock limited dredging effectiveness. Limited control over backfilling created higher than expected cap concentrations. No significant change in surface concentrations. Significant increase in some biota followed dredging.

Site: Lake Jarnson, Sweden

Stated Remedial Action Objectives (Related to Sediment Removal): Not a CERCLA remedy. Remediation goal was to substantially reduce transport of PCBs from lake sediments to lake water and downstream system to reduce PCB concentrations in biota (Fox River Group 1999).

Stated Cleanup Levels: PCBs at maximum of 0.5 ppm and no more than 25% of remediated area at over 0.2 ppm (Bremle et al 1998b).

Dates of Remediation: 1993-1994.

Were Cleanup Levels Achieved? Cleanup levels achieved except in one location.[25]

---

[25]After dredging was completed, 2.9 kg of PCB was calculated to be left in lake sediment. Sedimentary pool of PCB was distributed, and only one of 54 defined subareas exceeded 0.5 mg/kg dry weight. At commencement of remediation, this had been set as highest acceptable level to be left in sediment. Only 20% of sediment areas ended up with PCB higher than 0.2 mg/kg dry weight, which was better than remediation objective of 25% (Bremle et al. 1998b).

Were Remedial Action Objectives Achieved? Results indicate reductions in transport[26] and fish-tissue concentrations.[27]

Comments and Lessons Learned: Ability to overdredge. Sediment concentration decline corresponds to water and fish-tissue concentration declines. Monitoring data seek to differentiate regional declines in background PCB concentrations from those resulting from remediation. Did not target or dredge near-shore PCBs, which are later implicated as a continuing source of PCBs for fish.[28]

Site: Manistique Harbor, MI

Stated Remedial Action Objectives (Related to Sediment Removal): Reduce PCB concentrations in fish and water in the Manistique River and Harbor to levels that would not present an unacceptable human health or ecologic risk and would allow elimination of existing fish-consumption advisories. Maintain harbor as a navigable waterway for commercial shipping, fishing boats, and recreational watercraft. In general, restore river and harbor areas for use by deeper-draft vessels. Minimize need for future remedial action in area after completion of a non-time-critical action. Implement actions that would best contribute to efficient performance of any future remedial actions in the area. Achieve compliance consistent with federal and state ARARs for site. Comply with risk-based objectives defined by TERRA, Inc., as part of the risk assessment. Reduce, as much as practicable, the release of PCBs associated with particles and dissolved in the water to Lake Michigan (EPA 2006a [Manistique River and Harbor Site, May 10, 2006]).

Stated Cleanup Levels: Initially, goal of action was to remove sediments with PCB concentrations greater than 10 ppm. Later, goal was modified to state that objective was 95% removal of total PCB mass and an average sediment concentration not over 10 ppm throughout sediment column (Weston 2002).

Dates of Remediation: 1995-2000.

---

[26]"After completed remedial activities in Lake Jarnsjon in spring 1995, PCB concentrations had decreased at the outlet of the lake. In addition, the PCB transport during high flow in the early spring was lower than the year before" (Bremle et al. 1998b).

[27]"Fish from all the locations in 1996 [after dredging] had lower PCB concentration than in 1991 [before dredging]. The most pronounced decrease was observed in the remediated lake, where levels in fish were halved. The main reason for the reduced levels was the remediation" (Bremle and Larsson 1998).

[28]"Since some of the contaminated sediment was left in the lake after remediation and this sediment was mainly located in the most shallow, littoral areas, these sites constituted a probable source of PCB to zooplankton and fish. PCB still remaining in littoral sediment was probably the cause for a recorded gradient of PCB in fish from the lake and downstream the river" (Bremle and Larsson 1998).

Were Cleanup Levels Achieved? Average cleanup concentration level was met; it is unclear whether mass-removal goal was met.[29]

Were Remedial Action Objectives Achieved? Progress toward remedial action objectives after deposition event.

Comments and Lessons Learned: Initial cleanup levels not met. Poor initial characterization of wood debris and bedrock issues resulted in incomplete dredging and a longer dredging time frame. Dredging caused an initial increase in surface concentrations and no decrease in fish concentrations. Deposition due to dam removal and sand placement led to decreased surface PCB concentrations.

Site: Reynolds Metals, St. Lawrence River, NY

Stated Remedial Action Objectives (Related to Sediment Removal): Prevent human and biota contact with contaminated sediments. Reduce or prevent human ingestion of fish caught from the St. Lawrence River. Reduce short-term effects on surface water and air expected as a result of remedial activities (EPA 2006c).

Stated Cleanup Levels: PCBs, 1 ppm; PAHs, 10 ppm; TDBF, 1 ppb (EPA 1993b).

Dates of Remediation: 2001.

Were Cleanup Levels Achieved? Cleanup levels for PCBs were met after capping;[30] PAH cleanup levels not met; work continues.[31]

Were Remedial Action Objectives Achieved? Remedial action objectives not met. PAH remedial activities are ongoing.

Comments and Lessons Learned: Residuals due to bedrock and dredging over boulders and cobbles. PCB concentrations used to indicate PAH contamination; however, there was a lack of concordance between PCB and PAH contamination.

---

[29]"The range of removal efficiency stretches from 82% to 97% depending upon the many assumptions made in the calculation of the residual mass of PCBs. The outcome is highly dependant upon the specific gravity assumed for the in-situ sediments along with sediment volume estimates. The best case estimate indicates that the objective of 95% removal of PCBs from the AOC may have been met while the worst-case estimate indicates that the objective may not have been met." (Weston 2002).

[30]"Despite extensive dredging of the St. Lawrence River, the cleanup goals of 1 mg/kg PCBs, 10 mg/kg PAHs, and 1 µg/kg TDBFs were not achievable in all areas. As a result, a 0.75-acre, 15 cell area, containing a range of PCB concentrations from 11.1 mg/kg PCBs to 120.457 mg/kg, was capped with the first layer of a three-layer cap to achieve the cleanup goal [for PCBs]. The remaining exposed sediments average 0.8 mg/kg PCBs within the remaining 255 cells (21 acres), which is below the cleanup goal" (EPA 2006c).

[31]"The remedial action activities in the remaining cells containing elevated levels of PAHs above the cleanup goal have not been fully implemented" (EPA 2006c).

Site: Marathon Battery, Hudson River, Cold Spring, NY

Stated Remedial Action Objectives (Related to Sediment Removal): Reduce cadmium in sediments to protect aquatic organisms and protect human health. Reduce the transport of suspended sediments from east and west foundry coves and the pier area (EPA 1989a).

Stated Cleanup Levels: Dredging to 1 ft.

Comments: According to the Record of Decision (EPA 1989a), "The data compiled for east foundry cove indicate that over 95% of the cadmium contamination is located in the upper layer (1 foot) of the sediments. Due to the nature of the dredging process, dredging to a specific action level (for example, 10, 100, or 250 mg/kg of cadmium) would be technically difficult, since these concentrations vary in the sediments by only a few inches of depth. Therefore, expectations are that by dredging the upper layer of contaminated sediments, 95% of the cadmium contamination will be removed. Following remediation, it is anticipated that cadmium concentrations would not exceed 10 mg/kg in most of the dredged areas. . . . Sediment samples at and beneath the cold spring pier will be collected, analyzed, and evaluated to ascertain whether this area is a source of cadmium contamination. If, based upon this analysis, these sediments are determined to be a source, these sediments will be dredged to a depth of one foot."

Dates of Remediation: 1993-1995.

Were Cleanup Levels Achieved? Cleanup levels met (dredging performance targets met).

Were Remedial Action Objectives Achieved? Exposure measures of remedial action objectives met.

Comments and Lessons Learned: Discrete contaminated layer. Low amounts of debris. Ability to overdredge.[32] Site has useful post-dredging verification sampling and long-term monitoring data on sediments and biota that indicate beneficial effect of remedial activity.

Site: New Bedford Harbor, MA

Stated Remedial Action Objectives (Related to Sediment Removal): Pilot-project objectives were to significantly reduce PCB migration from hot-spot area sediment, which acts as a PCB source to the water column and to the remainder of the sediments in the harbor; to significantly reduce the amount of remaining PCB contamination that would need to be remediated to achieve overall harbor cleanup; to protect public health by preventing direct contact with hot-spot

---

[32]"EFC [East Foundry Cove] and EFP [East Foundry Pond] bottom sediments consist of silts and clays with some sand. These sediments have a very low bearing capacity which extend to nearly 80 feet in depth. The river bottom sediments consist of silts and clays with varying amounts of sand" (EPA 2006a [Marathon Battery Superfund Site, May 10, 2006]).

sediments; and to protect marine life by preventing direct contact with hot-spot area sediments.

Stated Cleanup Levels: Short-term hot-spot goal, 4,000 ppm total PCBs (EPA 1990a); long-term hot-spot goal, 10 ppm total PCBs after additional remediation occurs (EPA 1998b).

Comments: Owing to limited scope of hot-spot dredging action, EPA did not expect to achieve standards or levels of control associated with final cleanup levels (such as FDA PCB tolerance for fish tissue and water quality criterion). However, the action was expected to comply with some ARARs, including compliance with RCRA facility regulations, Executive Order 11988 regarding protection of flood plains to the extent practicable, Executive Order 11990 regarding protection of wetlands, and federal and state air standards during dredging and treatment of contaminated sediments (EPA 1990a).

Dates of Remediation: Hot-spot removal, 1994-1995.

Were Cleanup Levels Achieved? Operations appear to have achieved interim total-PCB cleanup level of 4,000 ppm (USACE 1995); removal was completed with minimal net transport of PCBs.

Were Remedial Action Objectives Achieved? Hot-spot removal was not intended to meet long-term risk-reduction goals. Unclear whether hot-spot removal objectives were met.

Comments and Lessons Learned: Usefulness of pilot studies. Value of EPA's process modifications on the basis of increased PCB and hydrogen sulfide concentrations in air during sediment handling. Indication of increased exposure to terrestrial plants and no increased exposure to aquatic biota during dredging. Challenge of relating contamination and dredging in a large harbor to human exposure and effects.

Site: United Heckathorn, Richmond, CA

Stated Remedial Action Objectives (Related to Sediment Removal): The ROD (EPA 1996b) states the clean up goal was based on a surface-water quality criterion for protection of human health from consumption of fish and bioaccumulation of DDT and dieldrin. DDT concentrations exceeded dieldrin concentrations by a factor of 10-100, so sediment remediation goals for both contaminants were based on DDT concentrations.

Stated Cleanup Levels: Water column, DDT at 0.59 ng/L and dieldrin at 0.14 ng/L. Sediment, DDT at 590 µg/kg (dry weight) (EPA 1996b).

Comments: Values are based on achieving a 10^{-6} lifetime excess cancer risk level. As described in the ROD (EPA 1996b), this value is lower than the chronic marine aquatic-life criterion of DDT at 1 ng/L or dieldrin at 1.9 ng/L. Human health criteria were judged likely to be achieved if average sediment DDT concentration was below 0.59 mg/kg (dry weight), which is lower than the 1 mg/kg that would probably meet marine chronic water quality criteria (EPA 1996b).

Dates of Remediation: 1996-1997.

Were Cleanup Levels Achieved? Cleanup levels apparently met immediately after dredging. Recontamination after dredging.

Were Remedial Action Objectives Achieved? Little improvement in channel; remedial action objectives not met.

Comments and Lessons Learned: Dredging was not effective in decreasing sediment or water concentrations, and biota concentrations did not decline to clean levels. Side slopes, piers, ship traffic, debris, and an outfall may have led to increased residual contamination. Difficulty in reaching agreement between parties. Conceptual site model was not sufficient to discern effect of dredging and likelihood of recontamination. Usefulness of deployed mussel studies to indicate effect of dredging.[33]

ABBREVIATIONS: PAH, polycyclic aromatic hydrocarbon(s); CERCLA, Comprehensive Environmental Response, Compensation, and Liability Act; PCB, polychlorinated biphenyl; SQO, sediment quality objective; ROD, record of decision; ESD, explanation of significant differences; CUL, cleanup level; BMP, best management practice; LDW, Lower Duwamish Waterway; COC, contaminants of concern; TSS, total suspended solids; TBT, tributyl tin; PRG, preliminary remediation goal; ARAR, applicable or relevant and appropriate requirements; TDBF, total dibenzofurans; RCRA, Resource Conservation and Recovery Act; DDT, dic.

---

[33]"Year 1 biomonitoring showed that pesticide concentrations in the tissues of mussels exposed at the site were higher than those observed before remediation. Year 2 samples, collected in February 1999, showed tissue levels that were much reduced from Year 1 but still exceeded pre-remediation levels of DDT at Lauritzen Channel, Santa Fe Channel, and Richmond Inner Harbor Channel" (EPA 2006a [United Heckathorn Superfund Site, Richmond, CA, May 16, 2006]).